I0064635

INTERACTION OF ELECTROMAGNETIC WAVES WITH ELECTRON BEAMS AND PLASMAS

INTERACTION OF ELECTROMAGNETIC WAVES WITH ELECTRON BEAMS AND PLASMAS

C S Liu
University of Maryland at College Park

V K Tripathi
Indian Istitute of Technology

World Scientific
Singapore • New Jersey • London • Hong Kong

Published by

World Scientific Publishing Co. Pte. Ltd.
5 Toh Tuck Link, Singapore 596224
USA office: 27 Warren Street, Suite 401-402, Hackensack, NJ 07601
UK office: 57 Shelton Street, Covent Garden, London WC2H 9HE

British Library Cataloguing-in-Publication Data
A catalogue record for this book is available from the British Library.

INTERACTION OF ELECTROMAGNETIC WAVES WITH ELECTRON BEAMS AND PLASMAS

Copyright © 1994 by World Scientific Publishing Co. Pte. Ltd.

All rights reserved. This book, or parts thereof, may not be reproduced in any form or by any means, electronic or mechanical, including photocopying, recording or any information storage and retrieval system now known or to be invented, without written permission from the publisher.

For photocopying of material in this volume, please pay a copying fee through the Copyright Clearance Center, Inc., 222 Rosewood Drive, Danvers, MA 01923, USA. In this case permission to photocopy is not required from the publisher.

ISBN-13 978-981-02-1577-4
ISBN-10 981-02-1577-0

PREFACE

During the last two decades, we have seen rapidly growing interest in the interaction of intense electromagnetic waves plasmas, primarily due to the availability of powerful lasers and other radiation sources, and their many applications, including laser fusion, beat wave accelerators, and heating of laboratory and ionospheric plasmas by radio waves. There has also been active research on new radiation sources such as gyrotrons and free electron lasers, employing relativistic electrons. Thus the study of electromagnetic wave interactions with plasma and electron beams is a rich, interdisciplinary field, with many diverse applications. There is, however, an underpinning of common theoretical concepts and mathematical techniques. In this book, we attempt to develop a self-contained and unified theoretical treatment of the basic processes of these interactions.

We have tried to elucidate the physics of these processes, and to develop the necessary mathematical methods of analysis. At high powers, the electromagnetic wave interaction with collective modes (waves) in plasmas becomes dominant, and we have paid special attention to the nonlinear mode-coupling processes such as parametric instabilities, and stimulated Raman and Brillouin scattering. In fact, these nonlinear processes qualitatively change the picture of interactions. According to the linear theory of electromagnetic wave propagation, the electromagnetic wave propagates in the underdense plasma, where the plasma frequency ω_p is less than the wave frequency ω, with dispersion and collisional damping, and is reflected at critical density where $\omega = \omega_p$. However, high-power electromagnetic waves can scattered predominantly backward and sideways by Raman and Brillouin processes, making the underdense region a scattering region. At critical density, the electromagnetic wave can parametrically decay into plasma waves and ion wave leading to anomalous absorption rather than reflection. Thus it is important to know the threshold power density and growth rate of these nonlinear processes.

In recent years excellent texts on waves in plasma and laser-plasma interactions have appeared. Kruer's book entitled *The Physics of Laser-Plasma Interactions* deals with many computer simulations and experiments related to laser-fusion research. Stix's new edition of *Waves in Plasmas* exhaustively covers linear and quasilinear properties of plasma waves, particularly waves in magnetized plasma. The present book is therefore limited in scope: only electromagnetic waves in unmagnetized plasmas or beams are considered, with emphasis on analytic approaches.

It can be used as a graduate text suitable for a one-semester course to introduce students to the physics of plasmas and electromagnetic waves in plasmas. It is also our hope that it may be used by physicists and engineers as a review of the fundamental processes of wave-plasma interactions because of the vast amount of literature dealing with this general topic.

CONTENTS

CHAPTER 1

INTRODUCTION

The interaction of electromagnetic waves with matter has always been a fascinating subject of study. As matter in the universe is mostly in the plasma state, the study of electromagnetic waves in plasma is of importance to astrophysics, space physics and ionospheric physics. For example, radio waves are not only important means of communication but also a useful tool for exploration of the ionosphere, magnetosphere, interstellar medium, pulsars and beyond.

With the advent of maser, laser, and other radiation sources, coherent electromagnetic waves, with well defined frequency and phase, have been shown to have wide ranging applications. Their interaction with plasma, in particular, has been a rich field of research in recent years. To explore the possibility of laser fusion,[1–4] laser-plasma interaction has been a subject of world wide research, revealing many novel nonlinear phenomena. To heat a magnetically confined plasma to the high temperature needed for fusion, radio waves of wide frequencies have been employed.[5] Ion cyclotron waves have been particularly successful for tokamak heating. Radio frequency waves are also used for driving dc current in a tokamak. Lower hybrid waves have been demonstrated to be efficient in this regard.

A plasma is an ionized state of matter with electrons and ions as its main constituents. A characteristic frequency of plasma is the frequency of electron plasma oscillation $\omega_p = (4\pi ne^2/m)^{1/2}$ where n, e and m are electronic density, charge and mass respectively. According to linear theory, when an electromagnetic wave of frequency ω is normally incident on a plasma, it is reflected if its frequency $\omega = \omega_p$. For waves of higher frequency

$\omega > \omega_p$ the plasma behaves as an optically rarer medium with refractive index $\eta = (1 - \omega_p^2/\omega^2)^{1/2}$. The collisions between the electrons and ions cause the wave to be damped.

The plasma also supports a low frequency ion acoustic wave with $\omega = kc_s(1 + k^2 c_s/\omega_{pi}^2)^{-1/2}$ and a high frequency Langmuir wave with $\omega = (\omega_p^2 + 3k^2 v_{th}^2/2)^{1/2}$, where c_s is the ion sound speed, v_{th} the electron thermal speed, k is the wave vector, and ω_{pi} is the plasma frequency. At shorter wavelengths both these waves are Landau damped by ions and electrons respectively.

This very simple theory serves as a useful guide for understanding the electromagnetic waves in plasma: their propagation in underdense plasma ($\omega_p < \omega$), absorption by collisions, reflection at critical density ($\omega_p = \omega$) and evanescence in the overdense plasma where $\omega_p > \omega$. But mode coupling becomes important for an obliquely incident electromagnetic wave onto an inhomogeneous plasma with electric polarization along the density gradient. Here the wave drives a density oscillation. At the critical density $n_c = (m\omega^2/4\pi e^2)$, the driven density oscillation is an electron plasma wave. Thus the electromagnetic wave energy can be directly converted to the electrostatic plasma wave through "resonance absorption", which eventually is absorbed by the plasma. This is a first example of the importance of collective modes for electromagnetic waves in plasma due to resonant mode coupling.

For a large-amplitude electromagnetic wave, its nonlinear interaction with other collective modes in plasma, through parametric instability, becomes dominant. At large amplitude, the electromagnetic wave can decay parametrically into a plasma wave and ion wave near the critical density. Energy of the electromagnetic wave is then diverted to these modes, which can heat the particles through Landau damping. In the underdense plasma, an electromagnetic wave can decay into another electromagnetic wave and an electrostatic wave, i.e., plasma (ion) wave, resulting in Raman (Brillouin) scattering. The nonlinear electromagnetic wave interaction with collective modes of plasma therefore qualitatively changes the nature of its propagation, absorption and scattering in plasmas.

There exist excellent texts treating many aspects of electromagnetic waves in plasma.[6-8] This monograph aims at reviewing the physical processes of linear and nonlinear collective interactions of electromagnetic waves with electron beams and unmagnetized plasmas. We have several specific applications in mind: coherent emission of high-power electromagnetic radiation from relativistic electron beams, advanced plasma accelerator, laser-driven-fusion, and high-power radiowaves in the ionosphere.

There is a great deal of interest in producing high-power coherent radiation in different frequency ranges from relativistic electron beams.[9-17] There are many kinds of plasma-coherent radiation sources, among them: Cerenkov free electron lasers (CFEL), cyclotron masers, and free electron lasers (FEL) are the prominent ones. In a CFEL[13] a relativistic electron beam propagates through a dielectric lined waveguide supporting a TM mode with finite axial component of the electric field. When the Doppler shifted frequency of the mode as seen by the beam is small, $\omega' \equiv \gamma_0(\omega - \mathbf{k} \cdot \mathbf{v}) \cong 0$, where γ_0 is the relativistic gamma factor, the mode appears as a dc field, resonantly exchanging energy with the electrons. As a continuous stream of electrons enters the interaction region in the presence of a TM mode of phase velocity slightly smaller than the velocity of the beam, half of the electrons are accelerated and the other half decelerated. The accelerated ones move quickly over to the phase, leading to net bunching of electrons in the retarding phase, hence, net amplification of coherent radiation.

This device has been successfully operated to produce radiation at 10.6 μm and longer wavelengths.

In a cyclotron maser[14-17] one employs a gyrating electron beam in a magnetic field in a cylindrical waveguide that supports TE and TM modes. In the absence of a radiation field, the electrons gyrate with energy-dependent cyclotron frequency, ω_c/γ_0 while gyrophases are uniformly distributed. In the presence of an electromagnetic wave of Doppler-shifted frequency $\omega' \equiv \omega - l\omega_c/\gamma_0 - k_z v_z \simeq 0$ (l being an integer) the electrons gain or lose energy depending on their gyrophase with respect to the wave. Those gaining energy become heavier, rotate slowly and phase-bunch with the ones losing energy and advancing in gyrophase. For $\omega' < 0$ this leads to coherent emission of radiation. The prominent devices based on the cyclotron maser instability are gyrotron, cusptron, peniotron and gyroklystron, which have been operated with high efficiency at millimetre wavelengths.

A free electron laser[9-11] is a fascinating device where a relativistic electron beam propagates through a periodic transverse magnetic field, called a wiggler. The wiggler appears to the beam as an incoming electromagnetic wave whose stimulated Compton backscattering produces double Doppler-shifted coherent radiation. In this process the wiggler and the radiation seed signal exert a ponderomotive force on the electrons causing space charge bunching, hence, stimulated emission of coherent radiation. The wavelength of the radiation goes directly as the wiggler period and inversely as the square of energy of the beam. The device is frequency tunable over a wide range. It operates at centimeter to micron wavelengths with high efficiency.

Laser-driven fusion is one of the great scientific challenges of our times. It aims at heating and compressing a D-T target by intense laser light to temperatures above 10 keV and superhigh-density when D-T nuclei would undergo an exothermic fusion reaction producing energy. Such high density is required for energy breakeven at 10 keV from the Lawson criterion which requires the product of density and confinement time $n\tau \geq 10^{14}$ cm^{-3}sec. For inertial confinement, τ is about the time required for the sound to travel a microsphere of radius of a few microns; thus the densities required are a thousand times higher than those in solids. This is achieved by shock compression of spherical D-T targets by the laser itself. The laser deposits its energy in the corona, i.e., the underdense region and the critical layer via collisional and resonance absorption. The heat is transported to the ablation layer by thermal conduction where a shock wave is generated that compresses the core.

The nonlinear interaction of intense laser light with the corona plasma is due to primarily parametric instability. The laser imparts a large oscillatory velocity to electrons that nonlinearly couple the collective modes of the plasma, giving rise to parametric instabilities.[18–20] In the underdense region, stimulated Brillouin, Raman and Compton scattering processes cause significant scattering of laser light. At the critical density the laser excites large amplitude Langmuir waves, through parametric decay into Langmuir wave and ion wave, which, besides heating the plasma, produce hot electrons that could preheat the core and diminish the efficiency of shock compression. The laser also has a tendency to break into filaments which may enhance Raman and Brillouin scattering. The Langmuir wave generated in linear mode conversion or parametric processes, however, is susceptible to modulational instability, generating solitons as well as collapsing, producing more hot electrons.

Plasma inhomogeneity plays an important role in these parametric wave-wave interactions. It limits the region of resonant wave-wave coupling, and enhances the threshold for parametric instabilities.

We discuss these processes in subsequent chapters. In Chapter 2, we discuss the properties of linear waves in a plasma. Starting from Maxwell's equations we discuss temporal and spatial dispersion and obtain expressions for energy density and energy flow in electromagnetic and electrostatic waves in a dispersive medium. We employ Vlasov theory for electrostatic waves whereas for electromagnetic waves a fluid description of plasma is considered appropriate. We discuss the phenomena of total internal reflection, diffraction divergence, dispersion broadening, surface wave propagation, duct propagation, Thomson scattering and anomalous resistivity. In Chapter 3, we study the process of linear mode conversion and resonance absorption in a warm plasma. In Chapter 4 the excitation of a plasma wave by two electromagnetic waves is developed and the concept of a beat wave accelerator is introduced. Two laser beams are taken to generate a large amplitude Langmuir wave at the difference frequency. The Langmuir wave can accelerate electrons to very high energies. Chapter 5 is devoted to coherent emission of radiation from electron beams. After discussing the physics of the bunching process we present linear theories of Cerenkov and undulator FEL. We also make gain estimates in the Compton regime. Chapter 6 deals with the self-focusing and filamentation of a high-powered electromagnetic beam in a plasma. There distinct cases are considered where nonlinearity arises on account of (1) relativistic mass variation, (2) ponderomotive force, and (3) differential Ohmic heating with and without thermal conduction. The wave equation is solved analytically in the WKB and paraxial ray approximations to study spatial evolution of beam radius and axial intensity. The problem of spatial growth of filamentation instability of a uniform electromagnetic wave has also been considered. In Chapter 7 we study parametric instabilities in a homogeneous plasma. We begin with the motion of a simple pendulum whose length is modulated by external means. The motion is governed by Mathiew's equation which we solve using perturbation theory. In the case of a parametric oscillator with two degrees of freedom we solve two coupled mode equations. Next we discuss the physics of three- and four-wave parametric processes in a plasma and solve coupled mode equations to obtain the growth rate. Towards the end we examine the parametric instability of a pump wave with random phase variation. Chapter 8 deals with the propagation and stability of Langmuir wave envelope solitons which are governed by a nonlinear Schrödinger equation. In Chapter 9 we study the effect of plasma inhomogeneity on parametric instabilities. For backscattering processes we employ a WKB theory giving a convective amplification factor. For side scattering as well as for parametric instabilities near the quarter critical and critical densities, we use full wave theory. Finally, we conclude by outlining the present status of research and future directions.

References

1. J. Nuckolls, L. Wood, A. Thiessen and Zimmerman, *Nature* **239**, 139 (1972).
2. N. G. Basov, V. A. Boiko, S. M. Zakhorov, D. N. Krokhin and G. V. Sklizkov, *JETP Lett.* **18**, 184 (1973).
3. C. S. Liu, *Bull. A. P. S.* **17**, 1065 (1972); C. S. Liu, M. N. Rosenbluth and R. B. White, *Phy. Fluid.*
4. W. L. Kruer, *The Physics of Laser Plasma Interactions*, (Addison-Wesley, 1987).
5. C. S. Liu and V. K. Tripathi, *Phys. Reports* **130**, 143 (1986).
6. T. Stix, *Waves Plasma*, (American Institute of Physics, 1992).
7. C. Bekef, *Radiation Processes in Plasma*, (Wiley, 1966).
8. V. L. Ginzburg, *The Propagation of Electromagnetic Wave in Plasmas*, (Pengamon, New York, 1971)
9. T. C. Marshall, *Free Electron Lasers*, (Macmillan, New York, 1985).
10. C. W. Roberson and P. Sprangle, *Phys. Fluids* **B1**, 3 (1989).
11. K. R. Chen, J. M. Dawson, A. T. Lin and T. Katsouleas, *Phys. Fluids* **B3**, 1270 (1991).
12. V. A. Flyagin, A. V. Gapanov, M. I. Petelin and V. K. Yulpatov, *IEEE Trans. Microwave Theory Tech.* MTT-25, 514 (1977).
13. J. E. Walsh, in "Free-electron generator of coherent radiation", *Phys. of Quant. Electron.* **7**, eds. S. F. Jacobs, H. S. Pilloff, M. Sargent III, M. O. Scully and R. Spitzer, (Addison-Wesley, 1980), p. 255.
14. P. Srangle, *J. Appl. Phys.* **47**, 2935 (1976).
15. K. R. Chu and J. L. Hershfield, *Phys. Fluids* **21**, 1877 (1987).
16. H. S. Uhm, R. C. Davidson and K. R. Chu, *Phys, Fluids* **21**, 1877 (1978).
17. D. Chernin and Y. Y. Lau, *Phys. Fluids* **27**, 2319 (1984).
18. K. Nishirawa, *J. Phys. Soc. Japan* **24**, 1152 (1968).
19. K. Nishikawa and C. S. Liu, in *Advances in Plasma Physics*, eds. A. Simon and W. B. Thompson, 6, 3 (1976); C. S. Liu and P. K. Kaw, *ibid*, p. 83; C. S. Liu, *ibid*, p. 121.
20. V. I. Karpman, *Nonlinear Waves in Dispersive Media*, (Novosibirsk Univ. Press, Novosibirsk, USSR, 1971) **14**, 40.

CHAPTER 2

BASIC EQUATIONS AND PROPERTIES OF LINEAR WAVES

2.1. Maxwell Equations

The Maxwell equations governing electromagnetic fields in a dielectric medium are

$$
\begin{aligned}
\nabla \cdot \mathbf{D} &= 4\pi\rho \,, \\
\nabla \cdot \mathbf{B} &= 0 \,, \\
\nabla \times \mathbf{E} &= \frac{1}{c}\frac{\partial \mathbf{B}}{\partial t} \,, \\
\nabla \times \mathbf{H} &= \frac{4\pi}{c}\mathbf{J} + \frac{1}{c}\frac{\partial \mathbf{D}}{\partial t} \,, \\
\mathbf{D} = \mathbf{E} + 4\pi\mathbf{P}, &\qquad \mathbf{B} = \mathbf{H} + 4\pi\mathbf{M} \,,
\end{aligned}
\tag{2.1}
$$

where c is the velocity of light in a vacuum, **E** and **H** are electric and magnetic fields, **D** and **B** are electric displacement and magnetic induction vectors, **P** and **M** are electric and magnetic dipole moments per unit volume caused by the displacement and orientation of bound electrons of neutrals atoms and molecules, and ρ and **J** are the charge and current densities due to free electrons and ions, related through the equation of continuity

$$\frac{\partial \rho}{\partial t} + \nabla \cdot \mathbf{J} = 0 \,. \tag{2.2}$$

In a fully ionized plasma $P = M = 0$. Quite commonly in plasma $P \ll E$ and $M \ll H$, hence, we shall assume in the following $\mathbf{D} = \mathbf{E}$ and $\mathbf{B} = \mathbf{H}$. The entire contribution of the plasma to the fields appears through $\mathbf{J}$, which must be obtained by solving kinetic or fluid equations, in terms of $\mathbf{E}$.

Wave equation

The wave equation governing the propagation of electromagnetic waves in a plasma can be obtained by taking curl of the third Maxwell equation and using the fourth,

$$\nabla^2\mathbf{E} - \nabla(\nabla \cdot \mathbf{E}) = \frac{4\pi}{c^2}\frac{\partial \mathbf{J}}{\partial t} + \frac{1}{c^2}\frac{\partial^2 \mathbf{E}}{\partial t^2} \; . \tag{2.3}$$

Current density and A.C. conductivity

In the limit of weak fields ($\mathbf{E} \to 0$), $\mathbf{J}$ is a linear function of $\mathbf{E}$; its value at $\mathbf{x}$, t, however, depends on the value of $\mathbf{E}$ not only at $\mathbf{x}$, t but also on $\mathbf{E}(\mathbf{x}', t')$ at other $\mathbf{x}'$ and $t' < t$ as the electric field affects the trajectories of the electron. Consider, for example, the response of an electron of mass m and charge $-e$ to an electric field $\mathbf{E}(t)$. The equation of motion

$$m\frac{d\mathbf{v}}{dt} = -e\mathbf{E} \tag{2.4}$$

leads to

$$\mathbf{v}(t) = -\frac{e}{m}\int_{-\infty}^{t} \mathbf{E}(t_1)dt_1 \; ,$$

$$\mathbf{J}(t) = -ne\mathbf{v} = -\frac{ne^2}{m}\int_0^{\infty} \mathbf{E}(t - t')dt' \; , \tag{2.5}$$

where n is the number density of electrons. When $\mathbf{E}$ is dependent on $\mathbf{x}(t)$ and t:

$$\begin{aligned} \mathbf{v}(t) &= -\frac{e}{m}\int_{-\infty}^{t} \mathbf{E}(\mathbf{x}(t_1), t_1)dt_1 \\ &= -\frac{e}{m}\iint_{-\infty}^{\infty}\iint_{-\infty}^{t} \delta(\mathbf{x}_1 - \mathbf{x}(t_1))\mathbf{E}(\mathbf{x}_1, t_1)d^3\mathbf{x}_1 dt_1 \end{aligned} \tag{2.6}$$

where $\mathbf{x}(t_1) = \mathbf{x}(t) + \int_t^{t_1} \mathbf{v}(t')dt'$. For small $\mathbf{E}$ the particle trajectory can be taken to be unmodified by the field, $\mathbf{x}(t_1) = \mathbf{x}(t) + \mathbf{v}_0(t_1 - t)$ where $\mathbf{v}_0$ is the unperturbed particle velocity and the integral can be explicitly evaluated.

The plasma response now explicitly depends on the spatial distribution of the field. This leads to the phenomenon of spatial dispersion.

In general the current density in a temporally and spatially dispersive medium can be written as

$$\mathbf{J}(\mathbf{x},t) = \iint_{-\infty}^{\infty} \iint_{0}^{\infty} \underline{\underline{\alpha}}(\mathbf{x}',t') \cdot \mathbf{E}(t-t',\mathbf{x}-\mathbf{x}')d^3\mathbf{x}'dt' \ . \tag{2.7}$$

For a monochromatic field,

$$\mathbf{E} = \mathbf{E}_0 e^{-i(\omega t - \mathbf{k}\cdot\mathbf{x})} \ , \tag{2.8}$$

Eq. (2.7) takes the form

$$\mathbf{J} = \underline{\underline{\sigma}} \cdot \mathbf{E} \ , \tag{2.9}$$

where

$$\underline{\underline{\sigma}}(k,\omega) = \iint_{-\infty}^{\infty} \iint_{0}^{\infty} \underline{\underline{\alpha}}(\mathbf{x}',t') e^{i(\omega t' - \mathbf{k}\cdot\mathbf{x}')d^3\mathbf{x}'dt'} \tag{2.10}$$

is the conductivity tensor. In an unmagnetized stationary plasma having no current, $\underline{\underline{\sigma}}$ acquires a simple form

$$\underline{\underline{\sigma}} = \sigma \underline{\underline{I}}; \qquad \underline{\underline{I}} = \begin{vmatrix} 1 & 0 & 0 \\ 0 & 1 & 0 \\ 0 & 0 & 1 \end{vmatrix} \tag{2.11}$$

with σ as the complex conductivity of the plasma.

2.2. Dispersion Relation

Substituting the plane wave solution (2.8) in the wave equation (2.3) we obtain

$$\left(k^2 - \frac{\omega^2}{c^2} - \frac{4\pi i \omega \sigma}{c^2}\right) \mathbf{E} = \mathbf{k}(\mathbf{k}\cdot\mathbf{E}) \ . \tag{2.12}$$

Multiplying Eq. (2.12) by $\mathbf{k}\cdot$ leads to

$$\varepsilon \mathbf{k}\cdot\mathbf{E} = 0 \ , \tag{2.13}$$

where

$$\varepsilon(\mathbf{R},\omega) \equiv 1 + \frac{4\pi i \sigma(\mathbf{R},\omega)}{\omega} \tag{2.13a}$$

is called the effective permittivity of the plasma. There are two possibilities of satisfying Eq. (2.13): (I) $\mathbf{k} \cdot \mathbf{E} = 0$, (II) $\varepsilon = 0$, giving electromagnetic and electrostatic waves respectively.

Electromagnetic waves

For $\mathbf{k} \cdot \mathbf{E} = 0$, Eq. (2.12) gives the dispersion relation

$$k^2 = \frac{\omega^2}{c^2}\varepsilon \, . \tag{2.14}$$

Further, from the third Maxwell's equation $\mathbf{B} = c\mathbf{k} \times \mathbf{E}/\omega$, hence, $\mathbf{E}$ and $\mathbf{B}$ are both perpendicular to $\mathbf{k}$. These are transverse electromagnetic waves.

Electrostatic waves

For $\mathbf{k} \cdot \mathbf{E} \neq 0$, one must have

$$\varepsilon = 0 \tag{2.15}$$

which on using in Eq. (2.12) yields $\mathbf{E}_{\|}\mathbf{k}$, giving $\mathbf{B} = c\mathbf{k} \times \mathbf{E}/\omega = 0$ and $\rho = (\sigma/\omega)\mathbf{k} \cdot \mathbf{E} \not\equiv 0$ (cf. Eq. (2.2)). This represents purely electrostatic space charge waves with $\mathbf{E}$ expressible in terms of a scalar potential ϕ

$$\mathbf{E} = -\nabla\phi \, . \tag{2.16}$$

In a cold plasma, there is no spatial dispersion, i.e., $\mathbf{J}$ at $\mathbf{x}$ depends only on the local value of $\mathbf{E}$ and not on its spatial distribution, then ε is independent of k and Eq. (2.15) gives the eigen-frequency of space charge oscillations. Since ω in this case is independent of $\mathbf{k}$ the group velocity of space charge waves ($\mathbf{v}_g = \partial\omega/\partial\mathbf{k}$) turns out to be zero.

One may view space charge oscillations a little differently too. Consider a uniform plasma of equal electron and ion densities n_0. The electronic charge and mass are $-e$ and m. Ions are assumed immobile. We produce a small space charge perturbation $\rho_1(\mathbf{x}, 0)$ at time $t = 0$ and wish to learn how $\rho_1(\mathbf{x}, t)$ would behave with time. ρ_1 produces an electric field $\mathbf{E}$ in the plasma that induces an electron current density $\mathbf{J}$. Multiplying Eq. (2.4) by $-n_0 e$ and approximating d/dt by $\partial/\partial t$ we get

$$\frac{\partial \mathbf{J}}{\partial t} = \frac{n_0 e^2}{m}\mathbf{E} \, .$$

This equation in conjunction with $\nabla \cdot \mathbf{E} = 4\pi\rho_1$ and $\partial\rho_1/\partial t + \nabla \cdot \mathbf{J} = 0$ yields

$$\frac{\partial^2 \rho_1}{\partial t^2} + \omega_p^2 \rho_1 = 0 \, ,$$

where $\omega_p = (4\pi n_0 e^2/m)^{1/2}$. The space charge thus oscillates at frequency ω_p. When collisional and thermal effects are included the oscillations die out with time. In a conductor where one could assume an instantaneous relationship between **J** and **E**, i.e., $\mathbf{J} = \sigma\mathbf{E}$ the space charge relaxation is governed by Eq. (2.2),

$$\frac{\partial \rho_1}{\partial t} + 4\pi\sigma\rho_1 = 0$$

which yields exponential decay of ρ_1 on a time $\sim (4\pi\sigma)^{-1}$.

2.3. Energy Density and Energy Flow

We return to Eqs. (2.1) to investigate the energy density of fields and the energy flow associated with a wave. Multiplying the fourth Maxwell equation by $\mathbf{E}\cdot$ and using vector identity $\nabla\cdot(\mathbf{E}\times\mathbf{H}) = \mathbf{H}\cdot\nabla\times\mathbf{E} - \mathbf{E}\cdot\nabla\times\mathbf{H}$

$$\mathbf{J}\cdot\mathbf{E} = -\frac{\mathbf{E}}{4\pi}\cdot\frac{\partial\mathbf{D}}{\partial t} - \frac{\mathbf{H}}{4\pi}\cdot\frac{\partial\mathbf{B}}{\partial t} - \frac{c}{4\pi}\nabla\cdot(\mathbf{E}\times\mathbf{H})\,. \tag{2.17}$$

In a nondispensive medium where **J** and **D** have an instantaneous relationship with **E**, as **B** has with **H**, energy densities of the electric and magnetic fields are given by $W_E = \mathbf{E}\cdot\mathbf{D}/8\pi$, $W_H = \mathbf{H}\cdot\mathbf{B}/8\pi$. Equation (2.17) has a simple interpretation: The power dissipation per unit volume (LHS) equals the rate at which field energy decreases plus the power flux that enters the unit volume (last term). The power flow density, called the Poynting vector, is thus

$$\mathbf{P} = \frac{c}{4\pi}\mathbf{E}\times\mathbf{H}\,. \tag{2.18}$$

In a plasma, however, **J** depends on the history of **E**, e.g., for sinusoidal fields **J** is almost $\pi/2$ out of phase with respect to **E** and $\mathbf{J}\cdot\mathbf{E}$ does not represent power dissipation. It is instead the rate of energy stored in the oscillatory motion of electrons. To determine this quantity, let us consider a field with slowly varying amplitude

$$\mathbf{E} = \mathbf{E}_0(t)e^{-i\omega t}\,. \tag{2.19}$$

Ignoring spatial dispersion the current density in an isotropic plasma can be written as (cf. Eq. (2.7))

$$\begin{aligned}\mathbf{J}(t) &= \int_0^\infty \alpha(t')\mathbf{E}(t-t')dt' \\ &\cong e^{-i\omega t}\int_0^\infty \alpha(t')\left[\mathbf{E}_0(t) - \frac{\partial \mathbf{E}_0}{\partial t}t'\right] e^{-i\omega t'}dt' \\ &= \left[\sigma \mathbf{E}_0 + i\frac{\partial \sigma}{\partial \omega}\frac{\partial \mathbf{E}_0}{\partial t}\right] e^{-i\omega t} ,\end{aligned} \tag{2.20}$$

where $\sigma(\omega) = \int \alpha(t')e^{i\omega t'}dt'$ is the conductivity of the plasma. In general $\sigma = \sigma_r + i\sigma_i$; for high frequency fields $\sigma_r \ll \sigma_i$. Using Eq. (2.20) we evaluate $\mathbf{J}\cdot\mathbf{E}$ averaged over $2\pi/\omega$

$$\begin{aligned}\mathbf{J}\cdot\mathbf{E}|_{\mathrm{av}} &= \frac{1}{2}\mathrm{Re}(\mathbf{J}\cdot\mathbf{E}^*) \\ &= \frac{1}{2}\sigma_r E_0^2 - \frac{1}{4}\frac{\partial \sigma_i}{\partial \omega}\frac{\partial E_0^2}{\partial t}\end{aligned} \tag{2.21}$$

and rewrite Eq. (2.17) as

$$\frac{1}{2}\sigma_r E_0^2 = -\frac{\partial}{\partial t}\overline{W}'_E - \frac{\partial}{\partial t}\overline{W}_H - \nabla\cdot\mathbf{P}_{\mathrm{av}} , \tag{2.22}$$

where $\mathbf{P}_{\mathrm{av}} = c\mathbf{E}\times\mathbf{H}^*/8\pi$ is the time-averaged Poynting flux, $\overline{W}_H = HH^*/16\pi$ is the time-averaged magnetic energy density, and

$$\begin{aligned}\overline{W}'_E &= \frac{1}{2}\frac{E_0^2}{8\pi}\left[1 - 4\pi\frac{\partial \sigma_i}{\partial \omega}\right] \\ &= \frac{1}{2}\frac{E_0^2}{8\pi}\left[\varepsilon_r + \omega\frac{\partial \varepsilon_r}{\partial \omega}\right]\end{aligned} \tag{2.23}$$

is the time-averaged energy density of the electric field plus the energy density of particles' motion. Here we have used Eq. (2.13a) for the definition of ε and employed the identity $\mathrm{Re}\,\mathbf{A}\cdot\mathrm{Re}\,\mathbf{B} = (1/2)\,\mathrm{Re}[\mathbf{A}\cdot\mathbf{B} + \mathbf{A}\cdot\mathbf{B}^*]$. One may note that W'_E is the same as the electric field energy density in a dispersive dielectric of permittivity ε_r.

In the case of electrostatic waves spatial dispersion is important. Consider a simple case

$$\begin{aligned}\mathbf{E} &= \hat{x}E_0(x,t)e^{-i(\omega t - kx)} \\ \mathbf{H} &= 0 .\end{aligned} \tag{2.24}$$

Defining conductivity $\sigma = \int \alpha(x',t')e^{i(\omega t' - kx')}dx'dt'$, one may write

$$\mathbf{J} \simeq \left[\sigma \mathbf{E}_0 - i\frac{\partial \sigma}{\partial k}\frac{\partial \mathbf{E}_0}{\partial x} + i\frac{\partial \sigma}{\partial \omega}\frac{\partial \mathbf{E}_0}{\partial t}\right] e^{-i(\omega t - kx)} ,$$

$$\frac{1}{2}\sigma_r E_0^2 = -\frac{\partial \overline{W}' E}{\partial t} + \frac{1}{16\pi}\frac{\partial E_0^2}{\partial x}\omega\frac{\partial \varepsilon_r}{\partial k} . \tag{2.25}$$

As we have seen in Sec. 2.2, $\varepsilon_r(\omega, k) \cong 0$ for electrostatic waves, one could write

$$\frac{\partial \varepsilon_r}{\partial k}\Delta k + \frac{\partial \varepsilon_r}{\partial \omega}\Delta \omega = 0 \tag{2.26}$$

or

$$\frac{\partial \varepsilon_r}{\partial k} = -\frac{\partial \varepsilon_r}{\partial \omega} V_g ,$$

where $V_g = \Delta\omega/\Delta k$ is the group velocity of the wave. Then the last term of Eq. (2.25), representing net power flux entering the unit volume, can be written as $-\partial \overline{P}/\partial x$, with

$$\overline{P} = \frac{1}{2}\omega\frac{\partial \varepsilon_r}{\partial \omega}\frac{E_0^2}{8\pi}V_g , \tag{2.27}$$

as the power flow density. One may recognize that $\overline{P}$ is the product of energy density (of the field and particle's drift motion) and the group velocity.

2.4. The Kinetic Equation

The particles in a plasma of finite temperature move with random velocities, each following a different trajectory. Since the response of charged particles to space-time dependent fields depends on the particles' trajectories it becomes necessary to adopt a statistical description. Define a particle distribution function $f(\mathbf{x}, \mathbf{p}, t)$ as the density of particles, of a given species, in six dimensional phase space $(\mathbf{x}, \mathbf{p})$. If collisions are ignored, f must satisfy an equation of continuity in six dimensional phase space

$$\frac{\partial f}{\partial t} + \frac{\partial}{\partial \mathbf{x}} \cdot (\dot{\mathbf{x}} f) + \frac{\partial}{\partial \mathbf{p}} \cdot (\dot{\mathbf{p}} f) = 0 . \tag{2.28}$$

Because $\mathbf{x}, \mathbf{p}$ are canonical conjugates of a Hamiltonian, $\partial \dot{\mathbf{x}}/\partial \mathbf{x} + \partial \dot{\mathbf{p}}/\partial \mathbf{p} = 0$ i.e., the flow in phase space is incompressible. Expressing $\dot{x} = \mathbf{v}$, $\dot{\mathbf{p}} = q(\mathbf{E} + \mathbf{v} \times \mathbf{B}/c)$ and realizing that $\mathbf{p}$ $(\equiv M\mathbf{v})$ and $\mathbf{x}$ are independent variables (with $q, M, \mathbf{v}$ and $\mathbf{p}$ as the particle's charge, mass, velocity and momentum), Eq. (2.28) becomes

$$\frac{\partial f}{\partial t} + \mathbf{v} \cdot \nabla f + q\left(\mathbf{E} + \frac{\mathbf{v} \times \mathbf{B}}{c}\right) \cdot \frac{\partial f}{\partial \mathbf{p}} = 0 . \tag{2.29}$$

It is known as the Vlasov equation. One may note from (2.29) that $df/dt = 0$, i.e., on a phase space trajectory, following a particle, f is a constant. In a non relativistic plasma one often defines $f(t, \mathbf{x}, \mathbf{v})$ as the particle density in $\mathbf{x}, \mathbf{v}$ space. However, Eq. (2.30) still holds good. When electron-ion Coulomb collisions are present, the Vlasov equation must be modified to include Coulomb collisions with the scatterers. This is the Fokker–Planck equation:

$$\frac{\partial f}{\partial t} + \mathbf{v} \cdot \nabla f + q(\mathbf{E} + \mathbf{v} \times \mathbf{B}/c) \cdot \frac{\partial f}{\partial \mathbf{p}} = \left.\frac{\partial f}{\partial t}\right|_{\text{coll.}} , \tag{2.30}$$

where the right hand side denotes the rate of change of the distribution function due to collisions.

Here, we do not evaluate the collision term and refer the reader to standard text books.[3,4] We defer the solution of the kinetic equation to a later section. We define charge and current densities in terms of f:

$$\begin{aligned} \rho &= \sum q \int f d^3\mathbf{v}, \\ \mathbf{J} &= \sum q \int \mathbf{v} f d^3\mathbf{v} , \end{aligned} \tag{2.31}$$

where the summation extends over electron and ion species.

In many cases of interest it suffices to adopt a fluid description of the plasma. The equations governing density, flow velocity and temperature of the fluid can be deduced from Eq. (2.30) by taking appropriate moments.

2.5. Fluid Equations

By definition the density n and average velocity $\langle \mathbf{v} \rangle$ of a species of a plasma are related to f through

$$\begin{aligned} n &= \int f d^3\mathbf{v} , \\ \langle \mathbf{v} \rangle &= \int f \mathbf{v} d^3\mathbf{v} . \end{aligned} \tag{2.32}$$

One could divide the velocity of particles into two components: a mean velocity $\langle \mathbf{v} \rangle$ and a random velocity $\mathbf{v}' = \mathbf{v} - \langle \mathbf{v} \rangle$. The average kinetic energy of random motion may be used to define the temperature

$$T = \frac{1}{3n}\int M(\mathbf{v} - \langle\mathbf{v}\rangle)^2 f d^3\mathbf{v}$$
$$= \frac{M}{3}\langle(\mathbf{v} - \langle\mathbf{v}\rangle)^2\rangle \ . \tag{2.33}$$

The equations describing the behavior of macroscopic parameters can be obtained by multiplying Eq. (2.30) by 1, $M\mathbf{v}$ and $Mv^2/2$ respectively and integrating over velocity. Following Braginskii[5] these equation can be cast into the following form:

$$\frac{\partial n}{\partial t} + \nabla\cdot(n\mathbf{v}) = 0 \ , \tag{2.34}$$

$$M\left(\frac{\partial \dot{\mathbf{v}}}{\partial t} + \dot{\mathbf{v}}\cdot\nabla\mathbf{v}\right) = \left(\mathbf{E} + \frac{1}{c}\mathbf{v}\times\mathbf{B}\right) - \frac{1}{n}\nabla(nT) - \mathbf{R} \ , \tag{2.35}$$

$$\frac{3}{2}\left(\frac{\partial T}{\partial t} + \mathbf{v}\cdot\nabla T\right) + T\nabla\cdot\mathbf{v} = q\mathbf{E}\cdot\mathbf{v} + \frac{1}{n}\nabla\cdot(\chi\nabla T) - Q \ , \tag{2.36}$$

where $\langle\mathbf{v}\rangle$ has been written as $\mathbf{v}$, for the sake of brevity, $\mathbf{R}$ is the mean momentum lost per second by a particle in collisions with other species, Q is the mean energy lost per second via elastic and inelastic collisions, χ is the thermal conductivity and we assumed that the velocity distribution function for the random velocity is isotropic.

Equation (2.34) is the equation of continuity for particle density, ignoring ionization and recombination processes. Equation (4.35), is the equation of motion of the fluid element. It states that the rate of change of average particle momentum, following the fluid element, equals the force due to electromagnetic fields, pressure gradient and collisional drag. In a fully ionized plasma collisional drag on electrons by ions is $\mathbf{R} = -m\nu(\mathbf{v}-\mathbf{v}_i)$ where m is the electron mass, ν is the electron-ion collision frequency and $\mathbf{v}$, $\mathbf{v}_i$ are electron and ion mean velocities.

Equation (2.36) is the energy balance equation. Its first two term S : $3dT/2dt$ represents, the rate of change of the particle's thermal energy following a fluid element, $-T\nabla\cdot\mathbf{v}$ denotes the rate of heat generated via fluid compression, $q\mathbf{E}\cdot\mathbf{v}$ is the power gained from the electric field and $-\frac{1}{n}\nabla\cdot(\chi\nabla T)$ is the power loss via thermal conduction. In the following we designate electron quantities as $M = m$, $q = -e$, $T = T_e$, $v_{\rm th} = (2T_e/m)^{1/2}$. For ions $M = m_i$, $q = ze$, $T = T_i$, $v_{\rm thi} = (2T_i/m_i)^{1/2}$, and the density and mean velocity are n_i, and $\mathbf{v}_i$. When dominant collisions of electrons are with ions, one could write $\mathbf{R}$ and Q for electrons as

$$\mathbf{R} = -m\nu(\mathbf{v} - \mathbf{v}_i) ,$$
$$Q = \frac{2m}{m_i}\nu(T_e - T_i), \tag{2.37}$$
$$\chi = \frac{nv_{th}^2}{\nu} ,$$

where

$$\nu = \nu_{ei} \equiv \left(\frac{2}{\pi}\right)^{1/2} \omega_p \frac{\ln(9n_D/z)}{9n_D/z} ,$$
$$\simeq 2.9 \times 10^{-6} nz \frac{\ln\Lambda}{T_e^{3/2}} \qquad (T_e \text{ in eV},\ n \text{ in cm}^{-3}) ,$$
$$n_D = \frac{4\pi}{3}\lambda_D^3 n = \frac{\pi\sqrt{2}}{3}\frac{v_{th}^3}{\omega_p^3} n = 1.7 \times 10^9 \left(\frac{T_e^3}{n}\right)^{1/2} ;$$
$$\omega_p = \left(4\pi ne^2/m\right)^{1/2} ,$$

ω_p and v_{th} are electron plasma frequency and thermal speed, $\lambda_D = v_{th}/\omega_p \sqrt{2}$ is the Debye length, and n_D is the number of electrons in a Debye sphere. For ions also one could write similar expressions. However, we shall defer writing $\mathbf{R}$, Q and χ for ions until necessary.

2.6. Plasma Response to an Electromagnetic Wave

Consider the propagation of an electromagnetic wave,

$$\mathbf{E} = \mathbf{E}_0 e^{-i(\omega t - \mathbf{k}\cdot\mathbf{x})} ,$$
$$\mathbf{B} = c\mathbf{k} \times \frac{\mathbf{E}}{\omega}, \qquad \mathbf{k}\cdot\mathbf{E} = 0 , \tag{2.38}$$

through a uniform plasma with no current. In the limit of $E_0 \to 0$, Eqs. (2.34)–(2.36) can be solved by expanding $n, \mathbf{v}, T$ around their equilibrium values and neglecting the products of perturbed quantities. Since $\mathbf{k}\perp\mathbf{E}$ it turns out that n, T perturbations vanish. Further, the ions' oscillatory velocity, due to their large mass, can be neglected as compared to that of the electrons. For electrons we obtain from Eq. (2.35)

$$\mathbf{v} = \frac{e\mathbf{E}}{mi(\omega + i\nu)},$$
$$\mathbf{J} = -n_0 e\mathbf{v} = \frac{n_0 e^2(\nu + i\omega)}{m(\omega^2 + \nu^2)}\mathbf{E} , \tag{2.39}$$

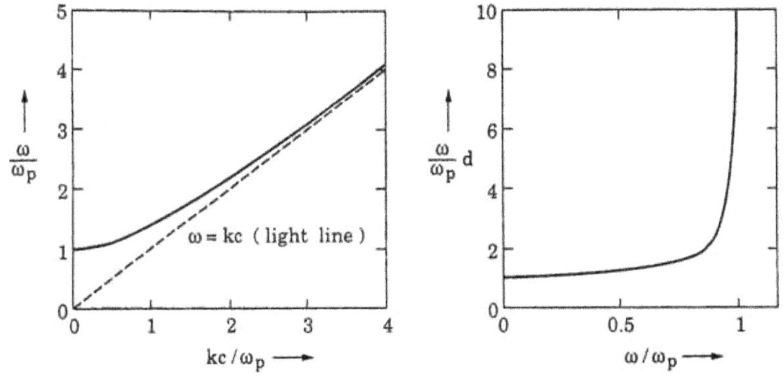

Fig. 2.1. Electromagnetic wave dispersion relation and depth of penetration $d = c(\omega_p^2 - \omega^2)^{-1/2}$ as a function of ω/ω_p.

giving complex conductivity $\sigma = -n_0e^2/im(\omega + i\nu)$ and effective permittivity

$$\varepsilon = 1 - \frac{\omega_p^2}{\omega^2 + \nu^2}\left(1 - i\frac{\nu}{\omega}\right) , \tag{2.40}$$

where n_0 is the electron density. Using ε in Eq. (2.14) one obtains

$$k^2 = \frac{\omega^2}{c^2}\left[1 - \frac{\omega_p^2(1 - i\nu/\omega)}{\omega^2 + \nu^2}\right] . \tag{2.41}$$

In the absence of collisions, Eq. (2.41) can be written as

$$\omega^2 = \omega_p^2 + k^2c^2 . \tag{2.42}$$

k is real only for $\omega > \omega_p$ (cf. Fig. 2.1). It vanishes at the critical density

$$n_{cr} = \frac{m\omega^2}{4\pi e^2} . \tag{2.43}$$

For $\omega < \omega_p$ (or $n_{cr} < n_0$), k is purely imaginary, hence, $\mathbf{B}$ is $\pi/2$ out of phase with $\mathbf{E}$ (cf. Eq. (2.38)) and the time average Poynting vector $\mathbf{P}_{av} = 0$, i.e., there is no energy propagation. The penetration depth of the field in the plasma is $c/(\omega_p^2 - \omega^2)^{1/2}$. For $\omega > \omega_p$ the refractive index of the plasma $\eta = (1 - \omega_p^2/\omega^2)^{1/2}$. For $\omega > \omega_p < 1$ and the product of phase and group velocities $v_{ph} \cdot v_g = c^2$. One may note from Eq. (2.49) that for real ω, k still could be complex, $\mathbf{k} = \mathbf{k}_r + i\mathbf{k}_i$ such that

$$\mathbf{k}_r \cdot \mathbf{k}_i = 0 ,$$

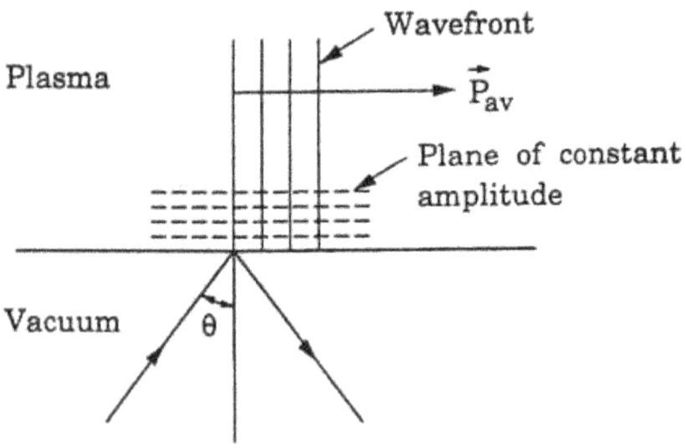

Fig. 2.2. Inhomogeneous electromagnetic wave in the plasma produced as a result of total internal reflection of an electromagnetic wave incident at an angle $\theta > \theta_c = \cos^{-1}(\omega_p/\omega)$.

i.e., the planes of constant phase and constant amplitude, defined by $\mathbf{k}_r \cdot \mathbf{x} =$ constant and $\mathbf{k}_i \cdot \mathbf{x} =$ constant respectively, are perpendicular to each other. These are inhomogeneous waves. Their time average Poynting vector turns out to be along $\mathbf{k}_r$. As an example consider a plane electromagnetic wave incident on a vacuum-plasma interface from the vacuum side (cf. Fig. 2.2). In vacuum, the components of $\mathbf{k}$ parallel and perpendicular to the interface are

$$k_{\parallel} = \frac{\omega}{c} \sin\theta$$

$$k_{\perp} = \frac{\omega}{c} \cos\theta$$

where θ is the angle of incidence. At the interface $k_{\parallel}$ is continuous (Snell's law), hence inside the plasma

$$\begin{aligned} k_{\perp} &= \sqrt{k^2 - k_{\parallel}^2} \\ &= \frac{1}{c}\left(\omega^2 \cos^2\theta - \omega_p^2\right)^{1/2} . \end{aligned}$$

For $\omega \cos\theta < \omega_p$, $k_{\perp}$ is imaginary, giving an inhomogeneous electromagnetic wave in the plasma. The energy flow in this case is along the interface.

The collisions introduce a component of current density in phase with the electric field, causing power dissipation and attenuation of waves. For $\omega > \omega_p, \nu$, Eq. (2.41) can be simplified to obtain $k \equiv k_r + ik_i$,

$$k_r = \frac{\omega}{c}\left(1 - \frac{\omega_p^2}{\omega^2}\right)^{1/2} ,$$
$$k_i = \frac{\nu}{2c}\frac{\omega_p^2}{\omega^2(1-\omega_p^2/\omega^2)^{1/2}} . \tag{2.45}$$

At very low frequencies $\omega < \nu$,

$$k \simeq \frac{1+i}{\sqrt{2}}\frac{\omega_p}{c}\left(\frac{\omega}{\nu}\right)^{1/2} .$$

The waves are strongly damped over a skin depth, $d_s = k_i^{-1} = (c/\omega_p)(2\nu/\omega)^{1/2}$. k_i is known as the attenuation constant. One might notice that the time-averaged Poynting vector $\mathbf{p} = c^2\mathbf{k}_r|E|^2/8\pi\omega$ decays like $\exp(-2\mathbf{k}_i \cdot \mathbf{x})$. Hence, power dissipation per unit volume is $-\nabla \cdot \mathbf{p} = 2k_r k_i c|E|^2/8\pi = \frac{1}{2}n_0\nu m|v|^2$, just the rate of dissipation of the electron oscillatory energy density due to the EM wave. The absorption efficiency is a function of $n_0\nu/n_{cr}$. In an inhomogeneous laser-produced plasma, a laser of higher frequency penetrates deeper to higher densities, hence, has a higher absorption efficiency. Further, slower electrons being more collisional are heated preferentially via collisional absorption, unlike the resonance absorption which produces hot electrons. Thus a short wavelength laser is preferred for laser driven fusion to avoid hot electrons. Experiments have reported absorption efficiency of $\gtrsim$ 80% for 0.35 μm light at 10^{15} W/cm^2 and moderate scale lengths.

A word would be in order here on the importance of kinetic effects that arise due to the random motions of electrons. They appear through (i) the velocity-dependent collision frequency ν, (ii) the Doppler shift in the frequency of the wave as seen by various particles and (iii) the $\mathbf{V} \times \mathbf{B}$ force. The latter two effects are significant when $\omega/k \sim v_{th}$ which is unlikely for electromagnetic waves in an unmagnetized non-relativistic plasma. The effect of velocity dependence of ν can be incorporated in the fluid equation if one employs a suitably averaged value of ν. It is useful to express electron oscillatory velocity in terms of average power flow density P:

$$\left|\frac{v}{c}\right| = \frac{e}{m\omega}\left(\frac{8\pi P}{c\eta}\right)^{1/2} \simeq 8\times 10^{-3}\lambda_\mu \frac{P_{14}^{1/2}}{\eta^{1/4}}$$

where P_{14} is the P expressed in units of 10^{14} W/cm^2, λ_μ is the free space wavelength in microns and η is the refractive index of the plasma. Thus for a Nd-glass laser with 1.06 μm wavelength and 10^{16} W/cm^2 intensity in an underdense plasma $\eta \sim 1$, $v/c \sim 0.1$ and the relativistic effect becomes important.

2.7. Electrostatic Waves

The space charge waves in a plasma are the compression–rarefaction waves of charged particles. Their phase velocity ranges from infinity to that somewhat greater than the electron thermal speed. Some particles of the plasma can be in resonance with them when the Doppler shifted frequency of the wave $\omega - \mathbf{k}\cdot\mathbf{v} \approx 0$ and they can be efficiently accelerated/decelerated by the wave. Consider for example an electrostatic wave $\mathbf{E} = -\hat{x}A\sin(\omega t - kx)$. In a frame moving with the phase velocity of the wave $\mathbf{E} = -\hat{x}A\sin k\Delta t$ and the x-component of electron velocity is $\Delta = v_x - \omega/k$. For resonant electrons, $\Delta \ll \omega/k$. We divide them into two groups: (A) $\Delta > 0$ and (B) $\Delta < 0$. Initially electrons of both the groups are uniformly distributed along the x-axis. Group A electrons ($v_x > \omega/k$) are accelerated in the accelerating zones (cf. Fig. 2.3), to move faster into the decelerating zones, and retarded in the decelerating zones to spend more time there. Thus there is bunching of Group A electrons in the retarding zones, transferring energy to the wave. The slower moving electrons ($v_x < \omega/k$) tend to bunch in the accelerating zones, gaining energy from the wave. The relative population of electrons with $v_x > \omega/k$ and $v_x < \omega/k$, i.e., the slope of the particle velocity distribution function at $v_x = \omega/k$, would decide net damping or growth of the wave. Thus it is necessary to employ kinetic theory to study the response of particles to electrostatic fields. We solve the Vlasov equation for the response of electrons in the presence of a Langmuir wave in a collisionless plasma. The electrostatic potential of the wave can be written as

$$\phi = \phi e^{-i(\omega t - \mathbf{k}\cdot\mathbf{x})} . \tag{2.46}$$

Expand the distribution function around its equilibrium value f_0,

$$f = f_0 + f_1$$

and linearize the Vlasov equation

$$\frac{\partial f_1}{\partial t} + \mathbf{v}\cdot\nabla f_1 = -\frac{e}{m}\nabla\phi\cdot\frac{\partial f_0}{\partial \mathbf{v}} \tag{2.47}$$

to obtain

$$f_1 = -\frac{e}{m}\int_{-\infty}^{t}\left(\nabla\phi\cdot\frac{\partial f_0}{\partial \mathbf{v}}\right)t'dt' , \tag{2.48}$$

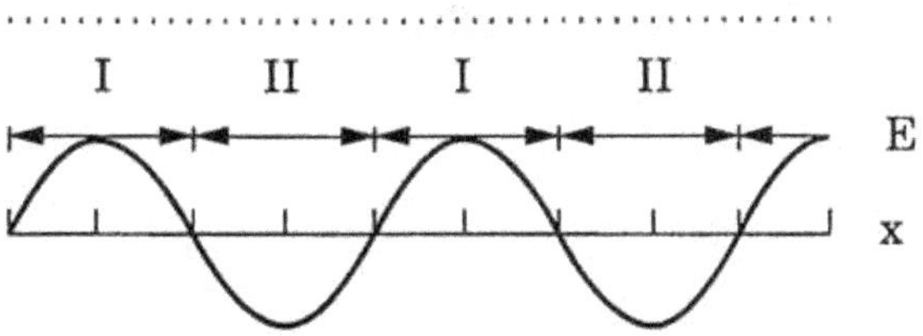

Fig. 2.3. Electric field of an electrostatic wave in a frame moving with the phase velocity of the wave. Zones I and II are respectively the retarding and accelerating zones for electrons moving towards $+x$.

where integration is over the unperturbed trajectory, $\mathbf{x}(t') = \mathbf{x}(t) + \mathbf{v}(t'-t)$. The integrand depends on the space-time history of the particle giving rise to spatial and temporal dispersion. Equation (2.48) can be simplified to give

$$f_1 = \frac{e\phi\mathbf{k}\cdot\frac{\partial f_0}{\partial\mathbf{v}}}{m(\omega - \mathbf{k}\cdot\mathbf{v})} . \tag{2.49}$$

For resonant particle the Doppler shifted frequency $\omega - \mathbf{k}\cdot\mathbf{v}$ vanishes and f_1 attains a large value. Integrating f_1 over velocity space, following Landau's prescription, one obtains the density perturbation

$$n_1 = \int f_1 d^3\mathbf{v} .$$

For a Maxwellian distribution function

$$f_0 = n_0\pi^{-3/2}v_{\rm th}^{-3}\exp[-v^2/v_{\rm th}^2] \tag{2.50}$$

n_1 takes the form

$$n_1 = \frac{k^2}{4\pi e}\chi_e\phi , \tag{2.51}$$

where

$$\chi_e = \frac{2\omega_{\rm p}^2}{k^2v_{\rm th}^2}\left[1 + \frac{\omega}{kv_{\rm th}}Z\left(\frac{\omega}{kv_{\rm th}}\right)\right] , \tag{2.52}$$

$$Z(\xi) = \frac{1}{\sqrt{\pi}}\int_{-\infty}^{\infty}\frac{e^{-x^2}}{x-\xi}dx ,$$

$\omega_p^2 = 4\pi n_0 e^2/m$, $v_{th}^2 = 2T_e/m$ and n and T_e are equilibrium electron density and temperature. The plasma dispersion function $Z(\xi)$ in the limiting cases of $\xi \ll 1$ and $\xi \gg 1$ can be expressed as

$$Z(\xi) \simeq -2\xi - \cdots + i\sqrt{\pi}e^{-\xi^2} \quad \text{for } \xi \ll 1 \, ,$$

$$Z(\xi) = -\frac{1}{\xi} - \frac{1}{2\xi^3} - \frac{3}{4\xi^5} - \cdots - i\sqrt{\pi}e^{-\xi^2} \quad \text{for } \xi \gg 1 \, . \tag{2.53}$$

Similarly the ion response can be written as

$$n_{1i} = -\frac{k^2}{4\pi ze}\chi_i\phi,$$

$$\chi_i = \frac{2\omega_{pi}^2}{k^2 v_{thi}^2}\left[1 + \frac{\omega}{kv_{th}} Z\left(\frac{\omega}{kv_{th}}\right)\right] \, , \tag{2.54}$$

where $\omega_{pi}^2 = 4\pi n_0 ze^2/m_i$ and $v_{thi}^2 = 2T_i/m_i$; n_0/z, ze, m_i and T_i are the equilibrium density, charge, mass and temperature of ions. Using the expressions for n_1 and n_{1i} in the Poissons equation

$$\nabla^2\phi = 4\pi e(n_1 - zen_{1i})$$

one obtains the dispersion relation for space charge modes

$$\varepsilon \equiv 1 + \chi_i + \chi_e = 0 \, . \tag{2.55}$$

Equation (2.55) has two distinct roots.

Langmuir waves

In the limit $\omega > kv_{th}$, kv_{thi}, i.e., when the distance travelled by particles in a wave period is shorter than a wavelength the ion susceptibility is m/m_i times smaller than the electron susceptibility and Eq. (2.55), on writing $\omega = \omega_r - i\Gamma$, yields

$$\omega_r^2 = \omega_p^2 + \frac{3}{2}k^2 v_{th}^2$$

$$\Gamma = \sqrt{\pi}\frac{\omega_p^3}{k^3 v_{th}^3}\omega_p \exp\left[\frac{-\omega_r^2}{k^2 v_{th}^2}\right] \, . \tag{2.56}$$

It is called a Langmuir wave. ω_r is always greater than ω_p. At short wavelengths as k approaches ω_p/v_{th} the Langmuir wave is strongly Landau damped (cf. Fig. 2.4). It may be noticed from Eq. (2.56) that the product of phase and group velocities is a constant:

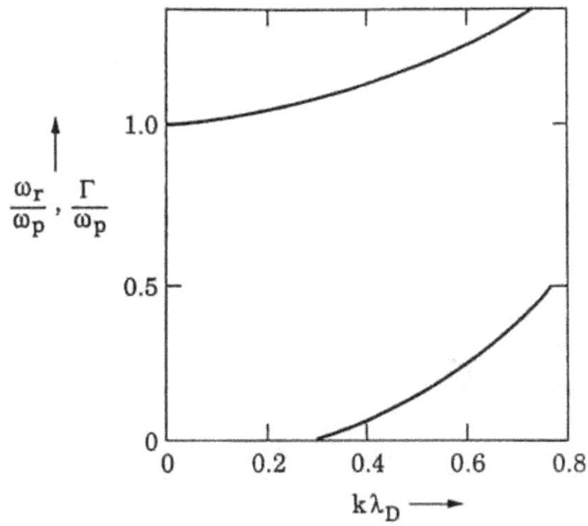

Fig. 2.4. Frequency (upper curve) and damping rate of a Langmuir wave as a function of wave number; $\lambda_D \equiv kv_{th}/\omega_p$.

$$V_g V_{ph} = \frac{3}{2} v_{th}^2 \,. \tag{2.57}$$

Ion acoustic wave

In the limit $kV_{thi} \ll \omega \ll kV_{th}$ Eq. (2.55) yields the ion acoustic wave: $\omega = \omega_r - i\Gamma$,

$$\omega_r = \frac{kc_s}{[1 + k^2c_s^2/\omega_{pi}^2]^{1/2}} \,,$$

$$\Gamma = \sqrt{\pi}\frac{\omega^3}{2k^2c_s^2}\left[\frac{\omega}{kv_{th}} + \frac{zT_e}{T_i}\frac{\omega}{kv_{thi}}e^{-(\omega/kv_{thi})^2}\right] \,, \tag{2.58}$$

where $c_s^2 = zT_e/m_i$. The maximum value of ω_r is ω_{pi}, the ion plasma frequency. However, at large values of k where $k \to \omega/v_{thi}$ the ion Landau damping is severe. Weak damping of these waves is possible only when $zT_e \gg T_i$ and $k^2c_s^2 \ll \omega_{pi}^2$. It may be noted from Eq. (2.51) that for ion acoustic waves the electron response is adiabatic $n_1 \simeq n_0 e\phi/T_e$ following Boltzman's law. In Fig. 2.5 we displayed the dispersion curve and the damping rate of an ion acoustic wave. One may recall that the damping of waves is caused by the pole of f_1, i.e., by the particles having $\omega \simeq \mathbf{k}\cdot\mathbf{v}$. When $\partial f_0/\partial v$ at $v = \omega/k$ is negative, i.e., more particles travel slower than the wave than those travelling faster than the wave, the wave becomes

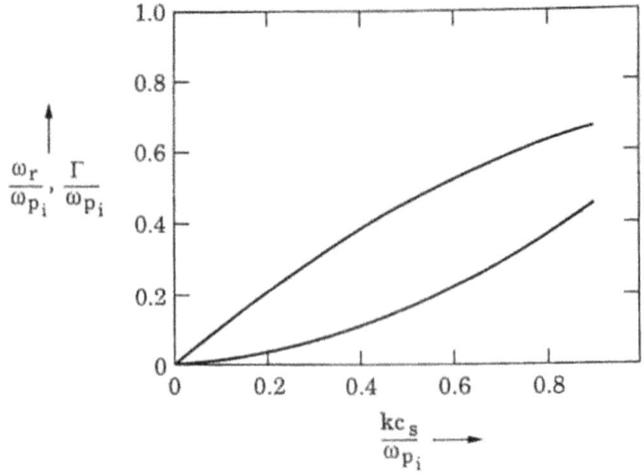

Fig. 2.5. Frequency (upper curve) and damping rate of an ion acoustic wave as a function of wave number.

damped because slower particles absorb energy from the wave and the faster particles give energy to the wave.

2.8. Diffraction Divergence

Now we consider electromagnetic waves of finite extent or duration propagating through plasmas and discuss effects of diffraction and dispersion. According to Huygen's principle every point of a wave front is a source of secondary wavelets, consequently a wave front of finite extent, i.e., non-uniform illumination, undergoes diffraction divergence. To have a quantitative estimate of this effect we consider a cylindrically symmetric electromagnetic beam propagating along $\hat{z}$ in a uniform plasma (cf. Fig. 2.6):

$$\mathbf{E} = \mathbf{E}_0(r \cdot z)e^{-i(\omega t - kz)}, \tag{2.59}$$

where

$$k = \frac{\omega}{c}\left(1 - \frac{\omega_p^2}{\omega^2}\right)^{1/2},$$

$$E_0(r, 0) = A_{00}e^{-r^2/2r_0^2},$$

$$\frac{\partial E_0}{\partial z} \ll kE_0 . \tag{2.60}$$

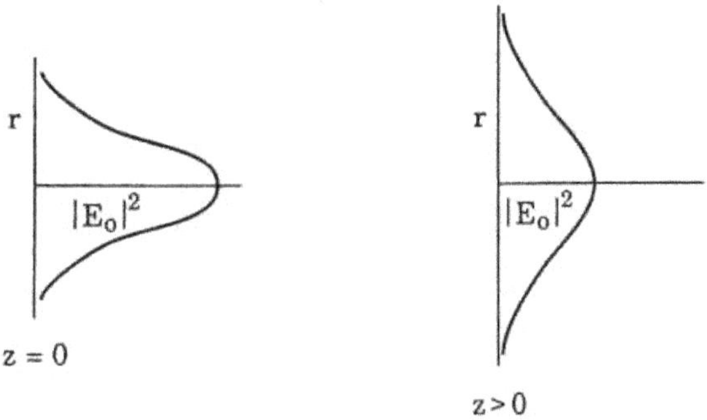

Fig. 2.6. Intensity profiles of an electromagnetic beam at $z = 0$ and $z > 0$. The size of the beam increases while intensity decreases due to diffraction effects.

Using Eq. (2.59) in the wave equation

$$\nabla^2 \mathbf{E} + \frac{\omega^2}{c^2}\left(1 - \frac{\omega_p^2}{\omega^2}\right)\mathbf{E} = 0 . \tag{2.61}$$

One obtains

$$2ik\frac{\partial E_0}{\partial z} + \nabla_\perp^2 E_0 = 0 . \tag{2.62}$$

Introducing an eikonal

$$E_0 = A_0(r,z)e^{iS(r,z)}$$

Eq. (2.62) on separating real and imaginary part yields

$$k\frac{\partial A_0^2}{\partial z} + \frac{\partial S}{\partial r}\frac{\partial A_0^2}{\partial r} + A_0^2 \nabla_\perp^2 S = 0 , \tag{2.63}$$

$$2k\frac{\partial S}{\partial z} + \left(\frac{\partial S}{\partial r}\right)^2 = \frac{1}{A_0}\nabla_\perp^2 A_0 , \tag{2.64}$$

where $\nabla_\perp^2 = \partial^2/\partial r^2 + (1/r)\partial/\partial r$. In compliance with the initial intensity distribution we construct the following Ansatz[6]

$$A_0 = \frac{A_{00}}{f(z)} e^{-r^2/2r_0^2 f^2(z)} \tag{2.65}$$

which conserves power flux as $\int A_0^2 r dr$ is independent of z. Substituting A_0 in Eqs. (2.63) and (2.64) we get,

$$\nabla_\perp^2 S - \frac{\partial S}{\partial r}\frac{2r}{r_0^2 f^2}\frac{2k}{f}\left(1 - \frac{r^2}{r_0^2 f^2}\right)\frac{df}{dz} = 0\,, \tag{2.66}$$

$$2k\frac{\partial S}{\partial z} + \left(\frac{\partial S}{\partial r}\right)^2 = -\frac{1}{r_0^2 f^2}\left(2 - \frac{r^2}{r_0^2 f^2}\right) \tag{2.67}$$

To solve Eqs. (2.66) and (2.67) in the paraxial region ($r^2 \ll r_0^2 f^2$) we expand S as

$$S = \phi(z) + \frac{\beta r^2}{2}\,. \tag{2.68}$$

Equating coefficients of various powers of r on both sides in the above equations we obtain

$$\beta = \frac{k}{f}\frac{df}{dz},$$

$$\frac{2k\partial\phi}{\partial z} = -\frac{2}{r_0^2 f^2}$$

and

$$\frac{d^2 f}{dz^2} = \frac{1}{R_d^2 f^3} \tag{2.69}$$

where $R_d = kr_0^2$. From the equation of the wave front, $kz + S =$ constant, one obtains the radius of curvature of the wave front on the axis $R = k/\beta$. For an initially plane wave front ($f = 1, df/dz = 0$ at $z = 0$) Eq. (2.75) yields

$$f^2 = 1 + \frac{z^2}{R_d^2}\,. \tag{2.70}$$

$R_d = kr_0^2$ is the characteristic length of diffraction divergence. The beam radius $r_0 f$ expands as the beam advances. The diffraction effect is less pronounced at shorter wavelengths, hence, these wavelengths are preferred for long distance point-to-point communication.

2.9. Dispersion Broadening

An electromagnetic pulse is a superposition of many frequencies. Since a plasma is a dispersive medium, different frequency components travel at different speeds resulting in pulse distortion. Consider the propagation of an initially gaussian (in time) electromagnetic pulse in a plasma

$$\mathbf{E} = \hat{x} E_0(z,t) e^{-i(\omega t - kz)} \tag{2.71}$$

with $E_0(0,t) = A_{00}e^{-t^2/\tau^2}$. One may recall that $\partial\mathbf{J}/\partial t = n_0e^2\mathbf{E}/m$, (cf. equation of motion) hence, on substituting (2.71) in the wave equation (2.3) we obtain

$$2i\omega\left(\frac{\partial E_0}{\partial t} + v_g\frac{\partial E_0}{\partial z}\right) - \frac{\partial^2 E_0}{\partial t^2} + c^2\frac{\partial^2 E_0}{\partial z^2} = 0 \tag{2.72}$$

where $v_g = c^2k/\omega = c\left(1 - \omega_p^2/\omega^2\right)^{1/2}$. For $\partial E_0/\partial t \ll \omega E_0$ the last two terms are small and one gets $\partial E_0/\partial t \simeq -v_g\partial E_0/\partial z$, i.e., $E_0(z,t) = F\left(t - z/v_g\right)$. To obtain a solution to higher order we approximate $\partial^2 E_0/\partial t^2 = v_g^2\partial^2 E_0/\partial z^2$ and introduce a new set of variables

$$z' = z\,,$$
$$t' = \left(t - \frac{z}{v_g}\right)\frac{\omega v_g}{\omega_p}\,.$$

Then Eq. (2.72) takes the form

$$2ik\frac{\partial E_0}{\partial z'} + \frac{\partial^2 E_0}{\partial t'^2} = 0 \tag{2.73}$$

which is the same as the wave in Eq. (2.62) $\nabla_\perp$ replaced by $\partial/\partial t'$. Following the procedure outlined in Sec. 2.9, we obtain

$$E_0 = A_0e^{iS}\,,$$
$$A_0^2 = \frac{A_{00}^2}{f(z')}e^{-t'^2/\tau'^2 f^2(z')}\,,$$
$$S = \psi(z') + \frac{1}{2}\beta(z')t'^2, \qquad \beta = \frac{k}{f}\frac{df}{dz'}\,,$$
$$\frac{d^2 f}{dz'^2} = \frac{1}{R_{dis}^2 f^2}\,, \tag{2.74}$$

where $\tau' = \tau c\omega/\omega_p$, $R_{dis} = k\tau'^2$. For $f(0) = 1$, $df/dz'|_{z'=0} = 0$, Eq. (2.74) yields

$$f^2 = 1 + \frac{z^2}{R_{dis}^2}\,. \tag{2.75}$$

The pulse undergoes dispersion broadening over a scale length $\sim R_{dis}$.

The dispersion effect has an interesting application in determining the distance of pulsars. Let two frequencies ω_1, ω_2 be generated simultaneously in an event in a pulsar. These two waves travel with group velocities $v_{g1} \simeq c\left[1 - \left(\omega_p^2/2\omega_1^2\right)\right]$ and $v_{g2} \simeq c\left[1 - \left(\omega_p^2/2\omega_2^2\right)\right]$, where ω_p is the plasma frequency of the interstellar medium. The time delay Δt between the two signals gives the distance L of the pulsar from the earth:

$$L \simeq \frac{c\Delta t \frac{\omega_1^2}{\omega_p^2}}{1 - \frac{\omega_1}{\omega_2}} .$$

2.10. Electromagnetic Wave Propagation in an Inhomogeneous Plasma

Let us now introduce one-dimensional inhomogeneity in the plasma (say $\nabla n_0 || \hat{x}$) and consider the propagation of a monochromatic wave at an angle to the density gradient in the x-z plane (cf. Fig. 2.7). For $\mathbf{E} \sim e^{-i\omega t}$ the wave equation (2.3) takes the form

$$\nabla^2 \mathbf{E} - \nabla(\nabla \cdot \mathbf{E}) + \frac{\omega^2}{c^2}\varepsilon \mathbf{E} = 0 , \qquad (2.76)$$

where $\nabla = (\hat{x}\partial/\partial x) + (\hat{z}\partial/\partial z), \varepsilon = 1 - [\omega_p^2(x)/\omega^2]$. The E_x and E_z equations are coupled but independent of E_y. There are two independent modes of propagation.

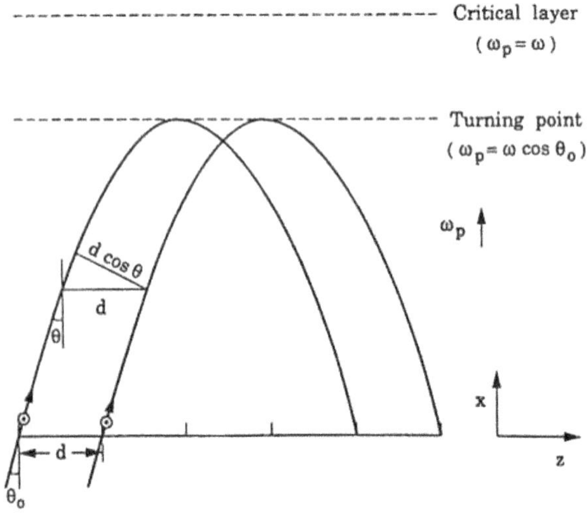

Fig. 2.7. Power flux tube in an inhomogeneous plasma where density increases with x. θ_0 is the angle of incidence at the entry point into the plasma.

(A) *S-Polarization*

$E_y \neq 0$; $E_x, E_z = 0$. In this case the y component of Eq. (2.76) is

$$\frac{\partial^2 E_y}{\partial x^2} + \frac{\partial^2 E_y}{\partial z^2} + \frac{\omega^2}{c^2}\varepsilon(x)E_y = 0 \, . \tag{2.77}$$

Expressing $E_y = f_1(x)f_2(z)$ Eq. (2.77) can be written as

$$\frac{1}{f_1}\frac{\partial^2 f_1}{\partial x^2} + \frac{\omega^2}{c^2}\varepsilon(x) = -\frac{1}{f_2}\frac{\partial^2 f_2}{\partial z^2} = k_z^2 \, , \tag{2.78}$$

where k_z^2 is an arbitrary constant, introduced because the LHS of (2.78) is independent of z and the RHS is independent of x, hence, the LHS and RHS both must be independent of x and z, i.e., a constant. For f_2 we obtain

$$f_2 = e^{ik_z z} \tag{2.79}$$

representing phase variation along the $+z$ direction. One could have also written an $e^{-ik_z z}$ term; however, we suppress it. The equation for f_1 can be written as

$$\frac{\partial^2 f_1}{\partial x^2} + k_x^2(x) f_1 = 0 \tag{2.80}$$

where $k_x^2 = (\omega^2/c^2)\varepsilon - k_z^2$. As long as the x variation of k_z^2 is slow, i.e., $\partial k_x / k_x^2 \partial x \ll 1$, we may write down a WKB solution

$$f_1 = \frac{E_0}{k_x^{1/2}} e^{i\int k_x dx} \, , \tag{2.81}$$

hence

$$\mathbf{E} = \hat{y}\frac{E_0}{k_x^{1/2}} e^{-i(\omega t - k_z z - \int k_x dx)} \, . \tag{2.82}$$

One may interpret k_x and k_z as the x and z components of the propagation vector. A very important characteristic of wave propagation in an inhomogeneous plasma is that k_z = constant, i.e., the component of the propagation vector transverse to the direction of inhomogeneity is constant.

The ray trajectory is given by

$$\frac{dx}{dz} = \frac{v_{gx}}{v_{gz}} \, . \tag{2.83}$$

Since $\mathbf{v}_g \equiv \partial\omega/\partial\mathbf{k} = c^2\mathbf{k}/\omega$, $v_{gx}/v_{gz} = k_x/k_z$, we have from Eq. (2.83)

$$z = k_z \int^x \frac{dx}{k_x(x)} + \text{const.} \tag{2.84}$$

Consider two rays, one passing through $x = 0$, $z = 0$, i.e., having

$$z_1 = k_z \int_0^x \frac{dx}{k_x(x)} \tag{2.85}$$

and the other having $x = 0$, $z = d$

$$z_2 = d + k_z \int_0^x \frac{dx}{k_x(x)} , \tag{2.86}$$

where subscripts 1 and 2 are just to identify the two rays cf. Fig. 2.7. It is clear from Eqs. (2.85) and (2.86) that at any value of x the z separation between the two rays is always d. The transverse separation between the two rays is $d\cos\theta = d \cdot k_x/k$ where θ is the angle a ray makes with the x axis. The power flux carried by a tube bound between these rays and having a width unity in the y direction is

$$\begin{aligned} P &= \left|\frac{c}{8\pi}\mathbf{E}^* \times \mathbf{H}\right| d\cos\theta = \frac{c^2}{8\pi\omega} k|E_y|^2 \frac{k_x}{kd} \\ &= \frac{c^2 d}{8\pi\omega} k_x |E_y|^2 = \text{ constant } . \end{aligned} \tag{2.87}$$

Thus the factor $k_x^{-1/2}$ in E_y ascertains that the power flux in any tube is constant, as long as absorption is unimportant.

One must notice that $k_x^2 = 0$ occurs where $\varepsilon = k_z^2 c^2/\omega^2$. If the wave is incident from vacuum making an angle θ_0 with the density gradient, then this condition implies

$$\omega_\mathrm{p} = \omega \cos\theta_0 . \tag{2.88}$$

Beyond this point k_x^2 is negative and the wave does not propagate. Equation (2.88) gives the turning point. However, near the turning point ($k_x \simeq 0$) the WKB approximation fails. If the density profile near the turning point is approximated to be linear

$$\omega_\mathrm{p}^2 = \omega_\mathrm{po}^2 \left(1 + \frac{x}{L_n}\right) , \tag{2.89}$$

Eq. (2.80) reduces to an Airy equation

$$\frac{d^2 f_1}{d\xi^2} - \xi f_1 = 0 , \tag{2.90}$$

giving the well behaved solution

$$f_1 = a_1 A_i(\xi) , \tag{2.91}$$

where

$$\xi = \frac{x - x_0}{\delta}, \qquad x_0 = \left(\frac{\omega^2 - \omega_{po}^2}{c^2} - \frac{\omega^2}{c^2}\sin^2\theta_0\right)\frac{c^2 L_n}{\omega_{po}^2}$$

$$\delta = (c^2 L_n \omega_{po}^2)^{1/3} ,$$

The scale length of the f_1 variation is δ. f_1 acquires large values (as compared to its values in the far under-dense region) near the turning point. For $\xi > 0$, the field decays rather rapidly with ξ.

(B) *P-Polarization*

$E_y = 0$; $E_x, E_z \neq 0$. For an electromagnetic wave polarized in the x-z plane $\nabla\cdot\mathbf{E}$ is not zero. From the equation of continuity $\rho = \nabla\cdot\mathbf{J}/i\omega = \nabla\cdot(\sigma\mathbf{E})/i\omega$ which, when employed in the first Maxwell equation, yields

$$\nabla\cdot(\varepsilon\mathbf{E}) = 0, \quad \text{i.e.,} \quad \nabla\cdot\mathbf{E} = -\frac{\mathbf{E}\cdot\nabla\varepsilon}{\varepsilon} .$$

Substituting for $\nabla\cdot\mathbf{E}$ in Eq. (2.76) we obtain for the x component

$$\nabla^2 E_x + \frac{d}{dx}\left(E_x \frac{d}{dx}\ln\varepsilon\right) + \frac{\omega^2}{c^2}\varepsilon E_x = 0 . \tag{2.92}$$

Following the method of separation of variables as outlined above we take the z variation of E_x as $e^{ik_z z}$. Introducing a new function

$$E_x = \frac{F(x)}{\varepsilon^{1/2}} e^{-i(\omega t - k_z z)} , \tag{2.93}$$

we obtain from Eq. (2.83)

$$\frac{d^2 F}{dx^2} + \left(\frac{\omega^2}{c^2}\varepsilon - k_z^2 - \frac{1 d}{2 dx}\left(\frac{1 d\varepsilon}{\varepsilon dx}\right) - \frac{1}{4}\left(\frac{1 d\varepsilon}{\varepsilon dx}\right)^2\right) F = 0 . \tag{2.94}$$

The turning point for the wave occurs at a density at which the big parenthesis of Eq. (2.94) vanishes. This does not happen at $\omega_p = \omega\cos\theta_0$ which is a turning point for the S-polarization. For long density scale lengths the two are not very different. Away from the turning point the last two terms in the big parenthesis can be dropped, then the KWB solution of Eq. (2.94) is the same as that of Eq. (2.80) for the S-polarized wave. Near the turning point and the singular point $\varepsilon = 0$ Eq. (2.94) needs to be reexamined carefully for inclusion of kinetic effects and the possibility of mode conversion to Langmuir waves. We defer this discussion to a separate chapter.

2.11. Surface Waves

At a sharp discontinuity in plasma density a new mode of propagation, called surface waves, exists. Consider a plasma-vacuum interfaces at $x = 0$. The effective permittivity is

$$\begin{aligned} \varepsilon = \varepsilon_{\mathrm{p}} &= 1 - \frac{\omega_{\mathrm{p}}^2}{\omega^2} \quad \text{for } x < 0 \\ &= 1 \quad \text{for } x > 0 \,. \end{aligned} \tag{2.95}$$

Away from the interface the wave equation for $E_z \sim e^{-i(\omega t - k_z z)}$ takes the form

$$\frac{d^2 E_z}{dx^2} - \alpha^2 E_z = 0 \,, \tag{2.96}$$

where $\alpha^2 = \alpha_{\mathrm{I}}^2 \equiv k_z^2 - \omega^2/c^2$ for $x < 0$ and $\alpha^2 = \alpha_{\mathrm{II}}^2 \equiv k_z^2 - (\omega^2/c^2)\varepsilon_{\mathrm{p}}$ for $x > 0$. A well behaved solution of the equation at $\pm\infty$ and continuous at $x = 0$ is

$$\begin{aligned} E_z &= A e^{\alpha_{\mathrm{I}} x} \quad \text{for } x < 0 \\ &= A e^{-\alpha_{\mathrm{II}} x} \quad \text{for } x > 0 \,. \end{aligned} \tag{2.97}$$

From the first Maxwell equation $E_x = -ik_z E_z/\alpha_{\mathrm{I}}$ for $x < 0$ and $E_x = ik_z E_z/\alpha_{\mathrm{II}}$ for $x > 0$. The continuity of εE_x at $x = 0$ gives the dispersion relation

$$\frac{\varepsilon_{\mathrm{p}}}{\alpha_{\mathrm{I}}} = -\frac{1}{\alpha_{\mathrm{II}}}$$

or

$$k_z^2 = \frac{\omega^2}{c^2} \frac{\varepsilon_{\mathrm{p}}}{\varepsilon_{\mathrm{p}} + 1} = \frac{\omega^2}{c^2} \frac{\omega_{\mathrm{p}}^2/\omega^2 - 1}{\omega_{\mathrm{p}}^2/\omega^2 - 2} \,, \tag{2.98}$$

giving a surface wave for $\omega < \omega_{\mathrm{p}}/\sqrt{2}$. The amplitude of the wave falls off away from the surface. Since the phase velocity ω/k_z of the wave is smaller than c it can be driven unstable by an electron beam propagating parallel to the interface.

Surface waves are an important mode of radio communication at medium frequencies. They propagate over the surface of the earth whose electrical conductivity σ is nearly real ($\sigma \sim 1$ mho/m) and $\varepsilon_{\mathrm{p}} \simeq 4\pi i\sigma/\omega \gg 1$. They travel hundreds of kilometers over the earth's surface without significant attenuation. Their field, however, decays rapidly inside the earth as one moves away from the interface. In vacuum the decay of field with x is less rapid.

2.12. Duct Propagation

A depressed density plasma duct has a tendency to trap and guide electromagnetic radiation, as observed in several experiments. Consider for example a density profile

$$\begin{aligned} n_0 &= n_{0\text{I}} \qquad \text{for } |x| < a \\ &= n_{0\text{II}} \qquad \text{for } |x| > a \end{aligned} \tag{2.99}$$

with $n_{0\text{I}} < n_{0\text{II}}$. The inner region has a higher refractive index than the outer. When a ray propagating in the inner region reaches the duct boundary at a large angle of incidence it is internally reflected back *in toto* and trapped inside the duct. Let us examine the propagation of a $\hat{y}$ polarized electromagnetic wave through the duct when $\partial/\partial y = 0$,

$$\mathbf{E} = \hat{y} E_0(x) e^{-i(\omega t - k_z z)} \tag{2.100}$$

where $E_0(x)$ is governed by the wave equation (2.80)

$$\frac{d^2 E_0}{dx^2} + k_x^2 E_0 = 0 \,, \tag{2.101}$$

$$\begin{aligned} k_x^2 = k_\text{I}^2 &\equiv \frac{\omega^2 - \omega_{\text{pI}}^2}{c^2} - k_z^2 \qquad \text{for } |x| < a \\ = -k_\text{II}^2 &\equiv \frac{\omega^2 - \omega_{\text{pII}}^2}{c^2} - k_z^2 \qquad \text{for } |x| > a \end{aligned} \tag{2.102}$$

with $\omega_{\text{pI, II}}^2 = 4\pi n_{0\text{I,II}} e^2/m$. The symmetric solution of Eq. (2.101) satisfying the continuity of E_y (the tangential electric field) at $x = \pm a$ and well behaved at $|x| \to \infty$ can be written as

$$\begin{aligned} E_0 &= A_1 \cos k_\text{I} x \qquad \text{for } |x| < a \\ &= A_1 \cos k_\text{I} a e^{-k_\text{II}(|x|-a)} \qquad \text{for } |x| > a \,. \end{aligned} \tag{2.103}$$

An integration of Eq. (2.101) across the discontinuity at $x = a$, from $a - \Delta$ to $a + \Delta$ with $\Delta \to 0$, yields dE_0/dx which is continuous at $x = a$. This condition leads to the dispersion relation

$$k_\text{I} a \tan k_\text{I} a = k_\text{II} a \equiv (\alpha^2 - k_\text{I}^2 a^2)^{1/2} \,, \tag{2.104}$$

where $\alpha^2 = (\omega_{\text{pII}}^2 - \omega_{\text{pI}}^2) a^2/c^2$. For a given α, Eq. (2.104) can be solved to obtain $k_\text{I} a$, which turns out to have many values, corresponding to different

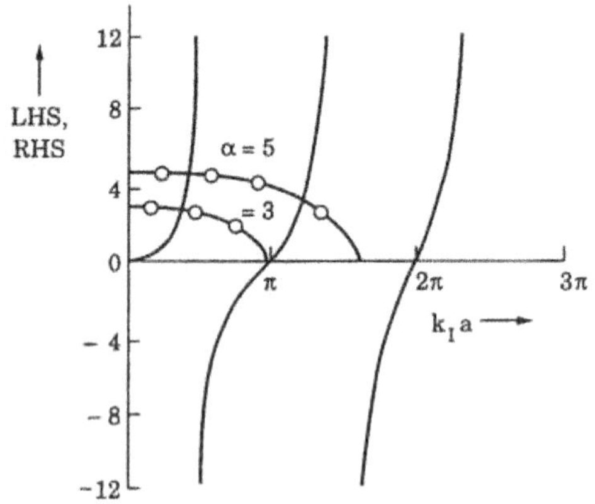

Fig. 2.8. LHS (——) and RHS (-0-0-0-) of the dispersion relation (2.104) as a function of $K_{\rm I}a$.

modes of propagation. In terms of a known $k_{\rm I}a$ and a the ω versus k_z relation for a mode can be straight away written from the definition (2.102)

$$\omega^2 = (k_{\rm I}^2 + k_z^2)c^2 + \omega_{\rm pI}^2 \,. \tag{2.105}$$

To illustrate the procedure of solution we have plotted in Fig. 2.8 the left and right hand sides (LHS and RHS) of Eq. (2.104) as a function of $k_{\rm I}a$ for a fixed α. The points of intersection of the two curves give the roots $k_{\rm I}a$ of Eq. (2.104). For $\alpha = 3$ only one root occurs, i.e., there is only one mode of propagation. For $\alpha = 5$ there exist two roots, i.e., two modes of propagation. The root with the smallest $k_{\rm I}a$ is called the fundamental mode. In Fig. 2.9 we have plotted the dispersion relation for the fundamental mode for $\alpha = 3$, $\omega_{\rm pI}a/c = 1$. The mode structure is plotted in Fig. 2.10. The mode extends to the outer region but falls off rapidly away from the duct boundary. There is a lower frequency cutoff for ducted propagation. For higher modes cutoff is higher. Asymptotically all modes approach $\omega = k_z c$.

This form of guided propagation of electromagnetic waves is important in a high-power free electron laser,[7,8] collective charged particle acceleration,[9] laser-driven fusion and ionospheric propagation.[10,11] In some applications it is more appropriate to model the density profile as parabolic,

$$n_0 = n_0^0 \left(1 + \frac{x^2}{L^2}\right) \,. \tag{2.106}$$

In that case the equation governing E_0 takes the form

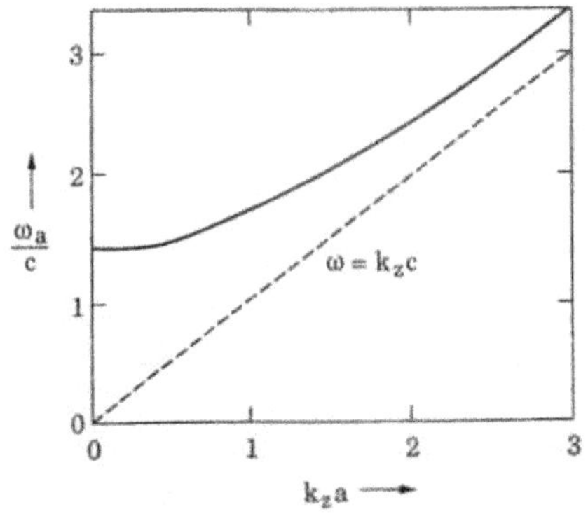

Fig. 2.9. Dispersion curve for the fundamental mode of duct propagation for $\alpha = 3$, $\omega_{pI}a/c = 1$ (i.e., $k_I a = 1.1$, $k_{II}a = 2.5$ determined self consistently).

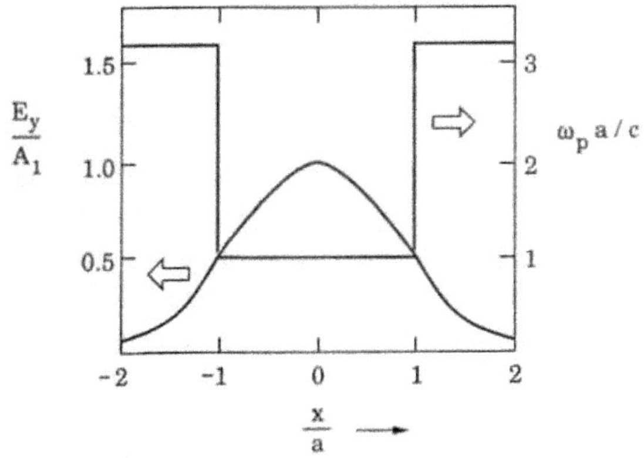

Fig. 2.10. Mode structure of the fundamental mode in a rectangular density duct.

$$\frac{d^2E_0}{d\xi^2} + (\mu - \xi^2)E_0 = 0\ , \tag{2.107}$$

giving

$$E_0 = A_n H_n(\xi) e^{-\xi^2/2}$$

and eigenvalues

$$\mu \equiv \left(\frac{\omega^2 - \omega_{po}^2}{c^2} - k_z^2\right)\frac{cL}{\omega_{po}} = n + \frac{1}{2}; \qquad n = 0, 1, \ldots\ , \tag{2.108}$$

where $\xi = x(\omega_{po}/cL)^{1/2}$, $\omega_{po}^2 = 4\pi n_0^0 e^2/m$, H_n are Hermite functions and

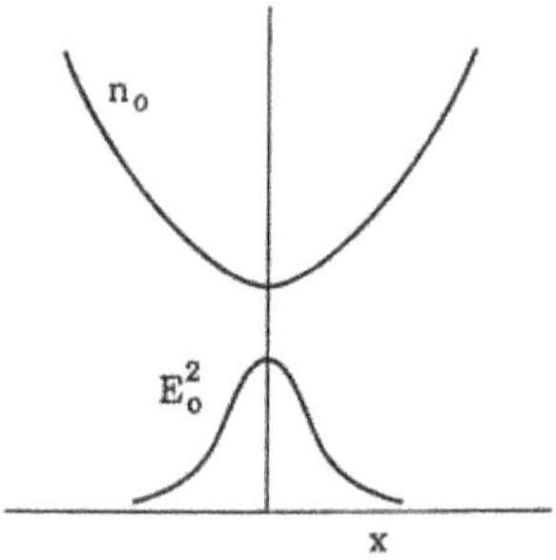

Fig. 2.11. Parabolic density profile and the mode structure of a trapped mode, $n = 0$.

A_n are constants. The fundamental ($n = 0$) mode has a gaussian distribution with half width $\delta \simeq (cL/\omega_{\rm po})^{1/2}$ (cf. Fig. 2.11).

2.13. Thomson Scattering of Radiation

In the presence of an electromagnetic wave the electrons of a plasma acquire oscillatory velocity and act as dipole sources of scattered radiation. If the electrons are at rest, the frequency of scattered radiation would be the same as that of the incident wave. However, for a moving electron the frequency of the incident wave is Doppler shifted. In a frame moving with the electron the scattered radiation is present at this Doppler shifted frequency. However, to a laboratory observer the frequency of scattered radiation is Doppler shifted a second time. The departure from the incident frequency is proportional to the electron velocity and depends on the angle of scattering. In a thermal plasma the electrons and ions have random distributions of velocities, hence, the scattered radiation has a broad line width.

Consider a plasma of volume V, having N electrons, moving randomly with position vectors $\mathbf{x}_1(t), \mathbf{x}_2(t), \ldots, \mathbf{x}_N(t)$. The electron density can be written as

$$n(\mathbf{x}, t) = \sum_j \delta(\mathbf{x} - \mathbf{x}_j(t)) \ . \tag{2.109}$$

The plasma is subjected to a small amplitude high frequency electromagnetic wave

$$\mathbf{E} = \mathbf{E}_0^{-i(\omega_0 t - \mathbf{k}_0 \cdot x)} \ . \tag{2.110}$$

Where $k_0 \simeq \omega_0/c$, $\omega_0 \gg \omega_{\rm p}$, $E_0 = E_{00}e^{-\gamma t}$ for $t > 0$ and $E_0 = 0$ for $t < 0$; $\gamma \to 0$. Under this field electrons are accelerated, $\mathbf{v} = -e\mathbf{E}/m$, and emit radiation. The far field of radiation at $\mathbf{R}, t$ due to an electron is[12]

$$\mathbf{E}_s(\mathbf{R},t) \simeq -\frac{e}{c^2}\frac{(\hat{q}\times\hat{q}\times\dot{\mathbf{v}})}{R}\Big|_{t'} , \tag{2.111}$$

where $\hat{q} = R/R$, and $t' = t - (1/c)|\mathbf{R} - \mathbf{x}(t')| \simeq t - (R/c) + (1/c)\hat{q}\cdot\mathbf{x}(t')$ is the advanced time at which an electron at position $\mathbf{x}(t')$ emits radiation. The total field due to all the electrons of the plasma is

$$\begin{aligned}\mathbf{E}_s(\mathbf{R},t) &= \frac{r_0}{R}(\hat{q}\times\hat{q}\times\mathbf{E}_0)\int d^3\mathbf{x}n(\mathbf{x},t')\cdot e^{-i(\omega_0 t' - \mathbf{k}_0\cdot x)} \\ &= \frac{r_0}{R}(\hat{q}\times\hat{q}\times\mathbf{E}_{00})\int_0^\infty\int dt' d^3\mathbf{x}n(x,t')e^{-i(\omega_0 t' - \mathbf{k}_0\cdot\mathbf{x})}e^{-\gamma t'} \\ &\cdot\delta\left(t' - t + \frac{R}{c} - \frac{\mathbf{q}\cdot\mathbf{x}}{c}\right)\end{aligned} \tag{2.112}$$

where $r_0 = e^2/mc^2$. Using the definition for a delta function

$$\delta(t) = \frac{1}{2\pi}\int_{-\infty}^{\infty} e^{i\omega t} d\omega . \tag{2.113}$$

One may write

$$\begin{aligned}\mathbf{E}_s(\mathbf{R},t) &= \frac{r_0}{R2\pi}(\hat{q}\times\hat{q}\times\mathbf{E}_{00})\iiint dt' d\omega d^3\mathbf{x}n(x,t') \\ &\cdot e^{-\gamma t'}e^{-i[(\omega_0-\omega)t' - (\mathbf{k}_0 - \hat{q}\frac{\omega}{c})\cdot\mathbf{x}]}e^{-i(\omega t - \frac{\omega}{c}R)} .\end{aligned} \tag{2.114}$$

Defining the Fourier–Laplace transform of n as

$$n(\mathbf{k},\omega) = \frac{1}{(\sqrt{2\pi})^3}\int_{-\infty}^{\infty}\int_0^\infty d^3\mathbf{x}dt' e^{i(\omega t' - \mathbf{k}\cdot\mathbf{x})}e^{-\gamma t'} n(\mathbf{x},t') ,$$

$\mathbf{E}_s(\mathbf{R},t)$ can be written as

$$\mathbf{E}_s(\mathbf{R},t) = \frac{1}{2\pi}\int_{-\infty}^{\infty} d\omega e^{-i\omega t}\mathbf{E}_s(\mathbf{R},\omega) \tag{2.115}$$

where

$$\begin{aligned}&\mathbf{E}_s(\mathbf{R},\omega) = \frac{r_0}{R}(\hat{q}\times\hat{q}\times\mathbf{E}_{00})(\sqrt{2\pi})^3 n(\mathbf{k}',\omega')e^{i\frac{\omega}{c}R} , \\ &\mathbf{k}' = \mathbf{k} - \mathbf{k}_0, \qquad \mathbf{k} = \frac{\omega}{c}\hat{q}, \qquad \omega' = \omega - \omega_0 .\end{aligned} \tag{2.116}$$

The total energy radiated per unit solid angle $W = R^2\int P dt$, where $P = c|\mathbf{E}_s(R,t)|^2/8\pi$ is the Poynting flux, can be written as

$$W = R^2 \frac{c}{8\pi} \int_{-\infty}^{\infty} |\mathbf{E}_s(R,\omega)|^2 \frac{d\omega}{2\pi}$$

where Parsevel's theorem has been used.

The quantity of interest is the ratio $(dW/d\omega)/W_{\text{in}}$ where $W_{\text{in}} = CE_{00}^2/16\pi\gamma$ is the total incident energy per unit area of the scaterring volume and $dW/d\omega$ is the energy radiated per unit solid angle per unit frequency interval,

$$\frac{1}{W_{\text{in}}} \frac{dW}{d\omega} = 2\sigma_{\text{T}} S(\mathbf{k}', \omega') \tag{2.117}$$

where $\sigma_{\text{T}} = 8\pi^3 r_0^2 (1 - \hat{q} \cdot \mathbf{E}_0 / |E_0|)^2$ is the incident power flux and $S(\mathbf{k}', \omega') = \gamma |n(\mathbf{k}', \omega')|^2$ is the spectral power density of electron density fluctuations.

Spectrum of density fluctuations $S(\mathbf{k}', \omega)$

Each charged particle in a plasma produces an electric field and influences the motion of all other particles. Thus there exists a finite correlation in the behavior of particles. To obtain the spectrum of density fluctuations we follow the dressed particle approach.[13–14] Consider an electron of charge $-e$ at position $\mathbf{x}_j(t)$ in a plasma. The density due to this electron

$$n_j(\mathbf{x}, t) = \delta(\mathbf{x} - x_j(t)) \;, \tag{2.118}$$

can be Fourier–Laplace transformed to give

$$n_j(\mathbf{k}, \omega) = \frac{1}{(\sqrt{2\pi})^3} g(\omega - \mathbf{k} \cdot \mathbf{v}_j) e^{i\mathbf{k} \cdot \mathbf{x}_j(0)} \tag{2.119}$$

where $g(\alpha) = \int_0^\infty e^{-\gamma t} e^{i\alpha t} dt$. Let the Fourier component of potential produced by this electron be $\phi(\mathbf{k}, \omega)$. Under this potential field the electrons and ions of the plasma acquire density perturbations

$$n'(\mathbf{k}, \omega) = \frac{k^2}{4\pi e} \chi_e(\mathbf{k}, \omega) \phi(\mathbf{k}, \omega)$$

$$n_i'(k, \omega) = -\frac{k^2}{4\pi e} \chi_i(\mathbf{k}, \omega) \phi(\mathbf{k}, \omega) \tag{2.120}$$

which, on using in Poisson's equation, yield

$$\phi(\mathbf{k}, \omega) = -\frac{4\pi e n_j(\mathbf{k}, \omega)}{k^2 \epsilon(\mathbf{k}, \omega)} \tag{2.121}$$

where χ_e and χ_i are electron and ion susceptibilities and $\varepsilon = 1 + \chi_e + \chi_i$. Each electron and ion of the plasma produces a similar potential. Taking electron and ion charges to be $-e$ and $+e$ respectively, the total potential due to all the charged particles can be written as

$$\phi(\mathbf{k},\omega) = -\frac{4\pi e}{k^2\varepsilon(k,\omega)}\frac{1}{(\sqrt{2\pi})^3} \times \left[\sum_{\substack{j \\ \text{electrons}}} g(\omega - \mathbf{k}\cdot\mathbf{v}_j)e^{i\mathbf{k}\cdot\mathbf{x}_j(0)} - \sum_{\substack{j \\ \text{ions}}} g(\omega - \mathbf{k}\cdot\mathbf{v}_j)e^{i\mathbf{k}\cdot\mathbf{x}_j(0)}\right] \tag{2.122}$$

and the electron density Fourier component as

$$\begin{aligned} n(\mathbf{k},\omega) &= n'(\mathbf{k},\omega) + \sum_{\substack{j \\ \text{electrons}}} n_j(\mathbf{k},\omega) \\ &= \frac{1+\chi_i(\mathbf{k},\omega)}{(\sqrt{2\pi})^3\varepsilon(\mathbf{k},\omega)} \sum_{\substack{j \\ \text{electrons}}} g(\omega - \mathbf{k}\cdot\mathbf{v}_j)e^{i\mathbf{k}\cdot\mathbf{x}_j(0)} \\ &\quad + \frac{\chi_e(\mathbf{k},\omega)}{(\sqrt{2\pi})^3\varepsilon(\mathbf{k},\omega)} \sum_{\substack{j \\ \text{ions}}} g(\omega - \mathbf{k}\cdot\mathbf{v}_j)e^{i\mathbf{k}\cdot\mathbf{x}_j(0)} \end{aligned} \tag{2.123}$$

One may now write $S(\mathbf{k},\omega) \equiv \gamma|n(\mathbf{k},\omega)|^2$, invoking the statistical independence of charged particles

$$\begin{aligned} S(\mathbf{k},\omega) &= \frac{\gamma}{2\pi}\left|\frac{1+\chi_i}{2\pi\varepsilon}\right|^2 \sum_{\substack{j \\ \text{electrons}}} |g(\omega - \mathbf{k}\cdot\mathbf{v}_j)|^2 \\ &\quad + \frac{\gamma}{2\pi}\left|\frac{\chi_e}{2\pi\varepsilon}\right|^2 \sum_{\substack{j \\ \text{ions}}} |g(\omega - \mathbf{k}\cdot\mathbf{v}_j)|^2 \end{aligned} \tag{2.124}$$

Using the definition of $g(\alpha) = 1/(\gamma - i\alpha)$ and replacing the summation by an integration over the velocity distribution functions, we obtain

$$\begin{aligned} S(\mathbf{k}',\omega') &= \frac{N}{8\pi^3}\left|\frac{1+\chi_i}{\varepsilon}\right|^2 \int_{-\infty}^{\infty} \frac{\gamma f_e(v_z)dv_z}{(\omega' - k'v_z)^2 + \gamma^2} \\ &\quad + \frac{N}{8\pi^3}\left|\frac{\chi_e}{\varepsilon}\right|^2 \int_{-\infty}^{\infty} \frac{\gamma f_i(v_z)dv_z}{(\omega' - k'v_z)^2 + \gamma^2} \end{aligned} \tag{2.125}$$

where $f_{e,i}$ are the one-dimensional distribution functions of electrons and ions normalized to unity, N is the total number of electrons in the scattering volume and the integrals are to be evaluated in the limit $\gamma \to 0$. When f_e, f_i are Maxwellian with temperatures T_e, T_i, $S(\mathbf{k}', \omega')$ simplifies to give

$$S(\mathbf{k}', \omega') = \frac{N}{8\pi^3}\left|\frac{1+\chi_i}{\varepsilon}\right|^2 \frac{1}{k' v_{th}\pi^{1/2}} e^{-(\omega'/k' v_{th})^2} + \frac{N}{8\pi^3}\left|\frac{\chi_e}{\varepsilon}\right|^2 \frac{1}{k' v_{thi}\pi^{1/2}} e^{-(\omega'/k' v_{thi})^2} \qquad (2.126)$$

where v_{th} and v_{thi} are the electron and ion thermal speeds, $\omega' = \omega - \omega_0$, $k' = |\mathbf{k} - \mathbf{k}_0|$. Since the scattered electromagnetic wave frequency $\omega \simeq \omega_0$, $k \sim k_0$, $k' \simeq 2k_0 \sin\theta_s/2$ where θ_s is the scattering angle. The expressions for electron and ion susceptibilities can be simplified in various frequency intervals as,

$$\omega' < k' v_{thi} \quad : \quad \chi_e \simeq \frac{2\omega_p^2}{k'^2 v_{th}^2}, \qquad \chi_i \simeq \frac{2\omega_p^2}{k'^2 v_{th}^2}\frac{T_e}{T_i}$$

$$k' v_{thi} < \omega' < k' v_{th} \quad : \quad \chi_e \simeq \frac{2\omega_p^2}{k'^2 v_{th}^2}, \qquad \chi_i \simeq -\frac{\omega_{pi}^2}{\omega'^2},$$

$$\omega' > k' v_{th} \quad : \quad \chi_e \simeq \frac{\omega_p^2 + \frac{3}{2}k'^2 v_{th}^2}{\omega'^2}, \qquad \chi_i \simeq 0 .$$

Consider two cases:

(i) $(2\omega_p^2/k'^2 v_{th}^2) \ll 1$, i.e., $\sin\left(\frac{\theta_s}{2}\right) \gg \frac{\omega_p}{\omega_0}\frac{c}{v_{th}\sqrt{2}}$

In this case both χ_e and χ_i are small and $S(k', \omega')$ takes the form

$$S(\mathbf{k}', \omega') = \frac{N}{8\pi^3 k' \pi^{1/2} v_{th}} \times \left[e^{-\omega'^2/k'^2 v_{th}^2} + \left(\frac{2\omega_p^2}{k'^2 v_{th}^2}\right)^2 \left(\frac{m_i}{m}\frac{T_e}{T_i}\right)^{1/2} e^{-\omega'^2/k'^2 v_{thi}^2}\right] . \qquad (2.127)$$

At very small values of $\sqrt{2}\omega_p/k' v_{th} < (mT_i/m_i T_e)^{1/2}$ the ion term is negligible and the spectrum of scattered radiation has a gaussian line-shape with frequency half width $\Delta \equiv k' v_{th} = 2\omega_0 \sin(\theta_s/2)$, which is just the Doppler broadening of scattered radiation by individual, unscreened electrons. The

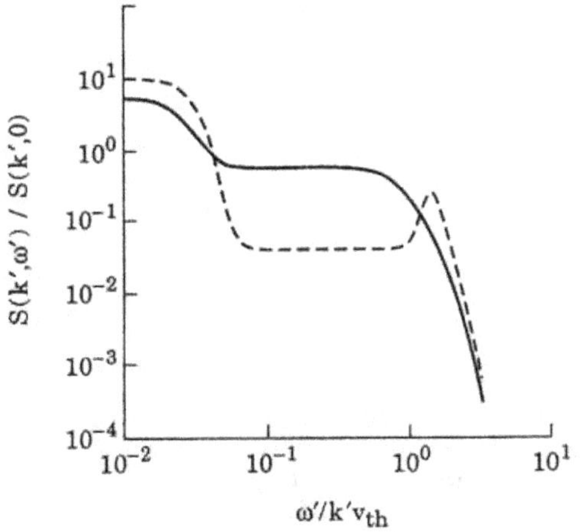

Fig. 2.12. Normalized spectral density as a function of $\omega'/k'v_{th}$ in a hydrogen plasma for $\sqrt{2}\omega_p/k'v_{th} = 0.5$ (-----) and 2 when $T_e = T_i$. [After Gerry and Patrick[15]]

collective effects are unimportant in this case. At smaller scattering angles $1 \gg \sqrt{2}\omega_p/k'v_{th} > (mT_i/m_iT_e)^{1/2}$ the second term in $S(\mathbf{k}',\omega')$ dominates the first one at small values of frequency shift $\omega' \lesssim k'v_{thi}$ and the line width is characterized by the ion thermal motion. At higher values of ω' the ion term is negligible.[15] Figure 2.12 shows the behavior of normalized spectral density $S(k',\omega')/S(k',_0)$ with $\omega'/k'v_{th}$ where,

$$S(\mathbf{k}',0) = \frac{N\left[\left(\beta^2 + \frac{T_e}{T_i}\right)^2 + \left(\frac{m_iT_e}{mT_i}\right)^{1/2}\right]}{8\pi^3 k'v_{th}\left(1 + \beta^2 + \frac{T_e}{T_i}\right)^2} \tag{2.128}$$

and $\beta = k'v_{th}/\sqrt{2\omega_p}$. The first transition occurs around $\omega' \sim k'v_{thi}$ at which the ion term falls and $S(\mathbf{k}',\omega')$ attains a constant value at intermediate values of ω'. As $\omega'/k'v_{th}$ approaches 1, the spectral density falls off rapidly as $\exp(-\omega'^2/k'^2v_{th}^2)$.

(ii) $\dfrac{2\omega_p^2}{k'^2v_{th}^2} > 1$

In this case χ_e and χ_i at low values of ω' are greater than 1 and the scattering is dominated by collective effects. At low values of $\omega' \lesssim k'v_{thi}$ the ion term (second in Eq. (2.126)) dominates the electron term, giving,

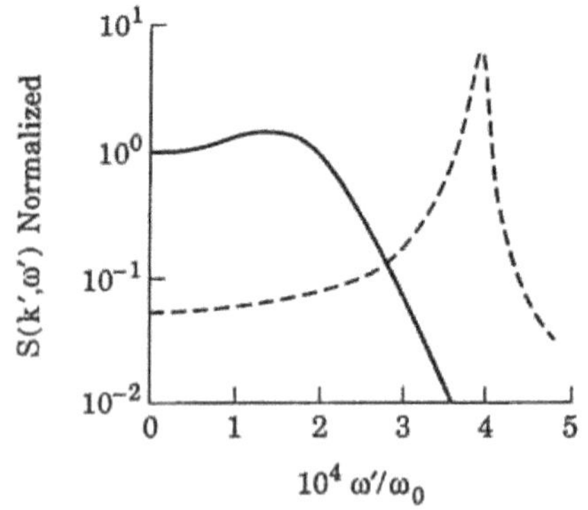

Fig. 2.13. Spectrum of scattered radiation in hydrogen plasma for $T_e/T_i = 1$ (——) and 10 (- - - - -), $k_0 = 3.76$ cm^{-1}, $\omega_p/\omega_0 = 0.5$, scattering angle 90°. [After Rosenbluth and Rostoker[16]]

$$S(\mathbf{k}',\omega') \simeq \frac{N}{8\pi^3 k' v_{\text{thi}}} e^{-\omega'^2/k'^2 v_{\text{thi}}^2} . \tag{2.129}$$

The line width is the characteristic of the ion thermal motion. At higher $\omega' > k' v_{\text{thi}}$ the ion term drops rapidly and $S(k',\omega')$ acquires an order of magnitude smaller value dominated by the electron thermal motions. Around $\omega' \sim \omega_p \left(1 + 3k^2 v_{\text{th}}^2/2\omega_p^2\right)^{1/2}$, ε takes very small values and the spectral density is reasonably enhanced (cf. Fig. 2.12). There also occurs a resonance near ion acoustic frequency $\omega' \sim k' c_s$ where c_s is the sound speed. However, it is important only in a non isothermal plasma with $T_e \gg T_i$. In an isothermal plasma with $T_e = T_i$, the ion Landau damping suppresses the ion acoustic resonance. Figure 2.13 displays the spectrum of scattered radiation[16] from a hydrogen plasma for different values of electron and ion temperatures. The electron plasma resonances are not shown in the figure as they occur at much higher values of $\omega' \equiv |\omega_s - \omega_0| \simeq \omega_p$.

The resonance at $\omega' \simeq \omega_p$, called the plasma line, is a powerful diagnostic tool. It occurs where $\varepsilon \simeq 0$. A relatively small departure of the velocity distribution function from the Maxwellian leads to orders of magnitude increase in the scattered power. It has been observed in the ionospheric backscatter experiments, and numerous laboratory experiments.

2.14. Anomalous Resistivity

The macroscopic behavior of a plasma is strongly influenced by the presence of density fluctuations. Let us apply a high frequency electric field $\mathbf{E}_{\omega_0}$

to a plasma having a certain level of density fluctuations $n_{\mathbf{k}}$. The field produces an oscillatory velocity $\mathbf{v}_{\omega_0}$ of electrons, resulting in a current density $-n_k e\mathbf{v}_{\omega_0}$, driving a resonant/non-resonant mode at $\omega_0, \mathbf{k}\pm\mathbf{k}_0$. The Landau damping of the driven mode causes dissipation of the high frequency field, giving rise to anomalous resistivity.[17]

Consider the propagation of an electromagnetic wave $\mathbf{E}_0 = \hat{x}A_0 e^{-i(\omega_0 t - \mathbf{k}_0\cdot\mathbf{x})}$ in a plasma having static turbulence, characterized by $\mathbf{k}$ spectrum of density fluctuations $n_{\mathbf{k}}$. The electrons acquire an oscillatory velocity $\mathbf{v}_0 = e\mathbf{E}_0/m_i\omega_0$ that beats with $n_{\mathbf{k}}$ to produce a nonlinear current and a nonlinear density perturbation

$$\mathbf{J}_1^{NL} = -\frac{1}{2}\sum_{\mathbf{k}} n_{\mathbf{k}} e\mathbf{v}_0 \,,$$

$$n_1^{NL} = \sum_{\mathbf{k}} \frac{n_{\mathbf{k}}\mathbf{k}\cdot\mathbf{v}_0}{2\omega_0} \,, \tag{2.130}$$

where we have used the equation of continuity and the fact that $\mathbf{k}_0\cdot\mathbf{E}_0 = 0$ for electromagnetic waves. In what follows we assume $k_0 \ll k$. The nonlinear perturbations produce a field $\mathbf{E}$ that we approximate to be electrostatic $\mathbf{E} = -\nabla\phi = -i\sum_{\mathbf{k}}\mathbf{k}\phi_{\mathbf{k}}$. Each $\phi_{\mathbf{k}}$ produces a self consistent linear density perturbation (cf. Eq. (2.51)):

$$n_{1\mathbf{k}}^L = \frac{k^2}{4\pi e}\chi_e(\omega_0,\mathbf{k})\phi_k \,. \tag{2.131}$$

Using $n_{1\mathbf{k}} = n_{1k}^L + n_{1k}^{NL}$ in the Poisson equation we obtain

$$\phi_k = -\frac{4\pi e}{k^2}\frac{\mathbf{k}\cdot\mathbf{v}_0 n_{\mathbf{k}}}{2i\omega_0\varepsilon(\omega_0,\mathbf{k})} \,, \tag{2.132}$$

where $\varepsilon(\omega_0,\mathbf{k}) = 1+\chi_e(\omega_0,\mathbf{k})$ is the plasma dielectric function. The electric potential $\phi_{\mathbf{k}}$ also produces an oscillatory velocity

$$\mathbf{v}_{1k} = -\frac{e\mathbf{k}\phi_{\mathbf{k}}}{m\omega_0}$$

which beats with $n_{\mathbf{k}}$ to produce a nonlinear current at $(\omega_0, 0)$:

$$\mathbf{J}_0^{NL} = -\frac{1}{2}\sum_{\mathbf{k}} n_k^* e\mathbf{v}_{1\mathbf{k}} = \frac{\omega_p^2}{4\omega_0^2}\sum_{\mathbf{k}}\frac{|n_{\mathbf{k}}|^2 e\mathbf{k}\cdot\mathbf{v}_0}{n_0\varepsilon(\omega_0,\mathbf{k})k^2}\mathbf{k} \,, \tag{2.133}$$

where n_0 is the equilibrium plasma density. Thus the total current at $(\omega_0, 0)$ is

$$\mathbf{J}_0 = -n_0 e \mathbf{v}_0 + \frac{\omega_p^2}{4\omega_0^2} \sum_{\mathbf{k}} \frac{|n_{\mathbf{k}}|^2 e \mathbf{k} \cdot \mathbf{v}_0}{n_0^2 \varepsilon(\omega_0, \mathbf{k}) k^2} \mathbf{k} \,. \tag{2.134}$$

For an isotropic turbulence

$$\mathbf{J}_0 = -n_0 e \mathbf{v}_0 \left(1 - \frac{\omega_p^2}{12\omega_0^2} \sum_{\mathbf{k}} \frac{|n_k|^2}{n_0^2 \varepsilon(\omega_0, k)}\right) \tag{2.135}$$

The ac conductivity can be written as

$$\sigma = -\frac{n_0 e^2}{m i \omega_0} \left[1 - \frac{\omega_p^2}{12\omega_0^2} \frac{1}{n_0^2} \sum_{\mathbf{k}} \frac{|n_{\mathbf{k}}|^2}{\varepsilon(\omega_0, \mathbf{k})}\right] \,. \tag{2.136}$$

The imaginary part of ε introduces an in-phase component of $\mathbf{J}_0$ with electric field $\mathbf{E}_0$. In cases when the turbulence spectrum is not discrete the summation in Eq. (2.136) must be replaced by an integration. In the case of a thermal level of fluctuations the conductivity reduces to $\sigma \simeq -(n_0 e^2/m i \omega_0)(1 - i\nu/\omega_0)$. However, when the fluctuation level is higher, the resistivity is considerably enhanced.

References

1. J. M. Dawson and C. Oberman, *Phys. Fluids* **5**, 517 (1962).
2. M. Rosenbluth and F. W. Perkins, Lecture Notes, Plasma Physics Laboratory, (Princeton, NJ, 1974).
3. I. P. Shkarofsky, T. W. Johnston and M. P. Bachincki, *The Particle Kinetics of Plasmas* (Addison-Wesley, 1966).
4. A. F. Alexandrov, L. S. Bogdankevich, and A. A. Rukhadze, in *Principles of Plasma Electrodynamics*, Springer-Verlag Series in Electrophysics **9** (1984).
5. S. I. Braginskii, *Sov. Phys. JETP* **6**, 358 (1958).
6. S. A. Akhmanov, A. P. Sukhorukov and R. V. Khokhlov, *Sov. Phys. Usp.* **10**, 609 (1968).
7. V. K. Tripathi and C. S. Liu, *IEEE Trans. Plasma Science* **18**, 466 (1990).
8. C. W. Roberson and P. Sprangle, *Phys. Fluids* **B1**, 3 (1989).
9. E. Esarey, A. Ting, and P. Sprangle, *Appl. Phys. Lett.* **53**, 1266 (1988).
10. M. S. Sodha, A. K. Ghatak, and V. K. Tripathi, *Progress in Optics* (North-Holland) **13**, 169 (1976).
11. F. W. Perkins and M. V. Goldman, *J. Geophys. Res.* **79**, 1478 (1974).
12. J. D. Jackson, *Classical Electrodynamics* (John Wiley, 1962)
13. M. N. Rosenbluth, *Topics in Advanced Plasma Theory* (Academic, New York, 1964).
14. G. Bekefi, *Radiation Processes in Plasmas* (John Wiley, 1966).
15. E. T. Gerry and R. M. Patrick, *Phys. Fluids* **8**, 208 (1965).
16. M. N. Rosenbluth and N. Rostoker, *Phys. Fluids* **5**, 776 (1962).
17. W. L. Kruer and J. M. Dawson, *Phys. Fluids* **15**, 446 (1972).

CHAPTER 3

RESONANCE ABSORPTION

An electromagnetic wave $\mathbf{E} = (\hat{x}E_{0x} + \hat{z}E_{0z})\exp[-i(\omega t - k_x x - k_z z)]$ obliquely incident onto a plasma with density gradient $\nabla n_0 \| \hat{x}$ produces an oscillatory current $\mathbf{J} = -n_0 e\mathbf{v}$ with $\nabla \cdot \mathbf{J} \neq 0$, causing space charge oscillations. At the critical layer, the frequency of oscillation would equal the natural frequency of plasma oscillation, hence, one might expect a resonant enhancement in its amplitude if the electromagnetic wave has an access to reach the critical layer.[1] The electromagnetic wave has a turning point ($k_x = 0$) at a lower density $n_0 = n_{cr}\cos^2\theta_0$ (where n_{cr} is the critical density and θ_0 is the angle of incidence at the plasma boundary, cf. Fig. 3.1), beyond which it is evanescent. Significant tunneling of electromagnetic energy through the evanescent region between the turning point and the critical layer is achieved when the width of this region is small, compared to a wavelength. In such a case a large amplitude Langmuir wave is excited near the critical layer. As it propagates towards smaller densities, it acquires large k and deposits energy on the electrons via Landau damping. This leads to strong absorption of radiation and production of hot electrons. In the following we follow the treatment of Piliya[2] to study linear mode conversion of an electromagnetic wave into a Langmuir wave, including thermal effects.

3.1. Current Density

Consider the propagation of a p-polarized electromagnetic wave in the x-z plane in a plasma with density gradient $\nabla n_0 \| \hat{x}$,

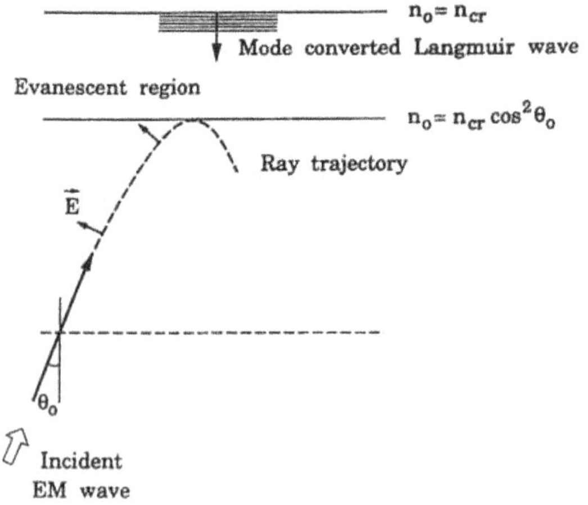

Fig. 3.1. Schematic of turning point and mode conversion layer.

$$\mathbf{E} = \mathbf{E}(x)e^{-i(\omega t - k_z z)} , \tag{3.1}$$

where $k_z = (\omega/c)\cos\theta_0$ and θ_0 is the angle of incidence at the entry point into the plasma. The linear response of electrons to this wave is obtained by solving the linearized equations of motion and continuity:

$$n = \frac{1}{i\omega}\nabla \cdot (n_0 \mathbf{v}) ,$$

$$\mathbf{v} = \frac{e\mathbf{E}}{mi\omega} + \frac{v_{\text{th}}^2}{i\omega}\frac{1}{n_0}\nabla n ,$$

$$\mathbf{J} = -ne\mathbf{v}$$

$$\simeq -\frac{n_0 e^2}{mi\omega}\left[\mathbf{E} - \frac{v_{\text{th}}^2}{\omega^2 n_0}\nabla\nabla \cdot (n_0\mathbf{E})\right] , \tag{3.2}$$

where higher order terms in v_{th}^2 have been dropped. Using Poisson's equation, $\nabla \cdot \mathbf{E} = -4\pi e n \simeq (4\pi e^2/\omega^2 m)\nabla \cdot (n_0\mathbf{E})$, the current density $\mathbf{J}$ can be rewritten as

$$\mathbf{J} = \frac{\omega}{4\pi i}\left[-\frac{\omega_{\text{p}}^2}{\omega^2}\mathbf{E} + \beta^2\frac{c^2}{\omega^2}\nabla\nabla \cdot \mathbf{E}\right] \tag{3.3}$$

where $\omega_{\text{p}}^2 = 4\pi n_0 e^2/m, \beta^2 = \frac{v_{\text{th}}^2}{c^2}$.

3.2. Coupled Mode Equations

The relevant Maxwell equation for the propagation of an s-polarized wave $(E_x, E_z, H_y \neq 0)$ are:

$$\nabla \times \mathbf{E} = \frac{i\omega}{c}\mathbf{H} \,, \tag{3.4}$$

$$\begin{aligned} \nabla \times \mathbf{H} &= \frac{4\pi}{c}\mathbf{J} - \frac{i\omega}{c}\mathbf{E} \\ &= -\frac{i\omega}{c}\left(\varepsilon\mathbf{E} + \frac{\beta^2 c^2}{\omega^2}\nabla\nabla\cdot\mathbf{E}\right) \,, \end{aligned} \tag{3.5}$$

where $\varepsilon = 1 - \omega_p^2(x)/\omega^2$. In component form these equations can be written as

$$ik_z E_x - \frac{\partial}{\partial x}E_z = \frac{i\omega}{c}H_y \,,$$

$$-ik_z H_y = -\frac{i\omega}{c}\varepsilon E_x + \frac{\beta^2 c^2}{\omega^2}\left(\frac{\partial^2}{\partial x^2}E_x + ik_z\frac{\partial}{\partial x}E_z\right) \,,$$

$$\frac{\partial}{\partial x}\left(H_y - \frac{ck_z\beta^2}{\omega}E_x\right) = -\frac{i\omega}{c}\left(\varepsilon - \beta^2\frac{k_z^2 c^2}{\omega^2}\right)E_z \,,$$

where we have replaced $\partial/\partial z$ by iR_z. Defining a new function

$$G(x) = H_y - \frac{k_z c}{\omega}\beta^2 E_x \,, \tag{3.6}$$

the above set can be written as

$$ik_z E_x - \frac{dE_z}{dx} = \frac{i\omega}{c}\left(G + \frac{k_z c}{\omega}\beta^2 E_x\right) \,, \tag{3.7}$$

$$-k_z G = -\frac{i\omega}{c}\varepsilon' E_x - \frac{i\omega}{c}\frac{\beta^2 c^2}{\omega^2}\frac{d^2 E_x}{dx^2} + \frac{k_z\omega}{c}\frac{\beta^2 c^2}{\omega^2}\frac{dE_z}{dx} \,, \tag{3.8}$$

$$\frac{dG}{dx} = -\frac{i\omega}{c}\varepsilon' E_z \,, \tag{3.9}$$

where $\varepsilon' = \varepsilon - k_z^2 c^2 \beta^2/\omega^2$. Using the value of dE_z/dx from Eq. (3.7) into (3.8) and that of E_z from (3.9) into (3.7) respectively we get

$$\beta^2 \frac{d^2 E_x}{dx^2} + \frac{\omega^2}{c^2}\left(\varepsilon' - \frac{k_z^2 c^2 \beta^2}{\omega^2}\right) E_x = \frac{k_z \omega}{c} G,$$

$$\frac{d^2 G}{dx^2} - \frac{d}{dx} \ln \varepsilon' \frac{dG}{dx} + \left[\frac{\omega^2}{c^2}\varepsilon' - k_z^2\right] G = -k_z^2 G + k_z \frac{\omega}{c}(1-\beta^2)\varepsilon' E_x \,.$$

When $k_z^2 v_{\text{th}}^2/\omega^2 \ll 1$ these equations can be simplified to give

$$\beta^2 \frac{d^2 E_x}{dx^2} + \frac{\omega^2}{c^2}\varepsilon E_x = \frac{k_z \omega}{c} G \,, \tag{3.10}$$

$$\frac{d^2 G}{dx^2} - \frac{1}{\varepsilon}\frac{d\varepsilon}{dx}\frac{dG}{dx} + \left(\frac{\omega^2}{c^2}\varepsilon - k_z^2\right) G = -\frac{ck_z}{\omega}\left[\frac{k_z\omega}{c} G - \frac{\omega^2}{c^2}\varepsilon E_x\right] . \tag{3.11}$$

3.3. Mode Conversion

Let us model the density profile as linear near the critical layer

$$n_0 = n_{0\text{cr}}\left(1 + \frac{x}{L_n}\right)$$

with $\frac{\omega}{c}L_n \gg 1$ and introduce a new variable $\xi = x/\lambda_{\text{em}}$ where $\lambda_{\text{em}} = (c^2 L_n/\omega^2)^{1/3}$ is called a characteristic electromagnetic wavelength. Equations (3.10) and (3.11) can now be written as

$$\frac{d^2 E_x}{d\xi^2} - \frac{c^2}{v_{\text{th}}^2}\xi E_x = \frac{k_z\omega}{c}\frac{c^2 \lambda_{\text{em}}^2}{v_{\text{th}}^2} G, \tag{3.12}$$

$$\frac{d^2 G}{d\xi^2} - \frac{1}{\xi}\frac{dG}{d\xi} - (\xi + k_z^2\lambda_{\text{em}}^2) G = -k_z^2 \lambda_{\text{em}}^2\left[G + \frac{\xi}{\lambda_{\text{em}}^2}\frac{c}{k_z\omega}E_x\right] . \tag{3.13}$$

We notice two different scale lengths for E_x and G, *viz*, $\xi \sim v_{\text{th}}/c$ and $\xi \sim 1$ respectively. Further, if $v_{\text{th}} \to 0$, Eq. (3.12) gives $E_x = -\lambda_{\text{em}}^2 k_z \omega G/c\xi$ and the right hand side of Eq. (3.13) identically vanishes. For $(c^2/v_{\text{th}}^2)\xi \gg 1$ one may write down an asymptotic (WKB) solution to Eq. (3.12)

$$E_x = -\frac{k_z\omega}{c}\lambda_{\text{em}}^2 \frac{G(\xi)}{\xi} + a_1 \frac{v_{\text{th}}}{c}(-\xi)^{-1/4} e^{i\frac{c}{v_{\text{th}}}\frac{2}{3}(-\xi)^{3/2}} \,, \tag{3.14}$$

where the first term is the particular solution corresponding to the transverse wave and the last term, with a constant a_1, is the complimentary solution representing a plasma wave propagating towards $-\hat{x}$. A plasma wave going along $+\hat{x}$ has been dropped as there is no source for these waves in the underdense region.

The WKB solution to Eq. (3.13), for $\xi \gg 1$, can be written as

$$G = b_1(-\xi - k_z^2\lambda_{\rm em}^2)^{-1/4}e^{-i2/3(-\xi-k_z^2\lambda_{\rm em}^2)^{3/2}}$$
$$+ b_2(-\xi - k_z^2\lambda_{\rm em}^2)^{-1/4}e^{+i2/3(-\xi-k_z^2\lambda_{\rm em}^2)^{3/2}} \quad . \tag{3.15}$$

We have dropped the particular integral which is $\approx a_1 v_{\rm th}^3/c^3(-\xi)^{+3/4}$ $(k_z c/\omega)$. The two terms in Eq. (3.15) represent incident and reflected electromagnetic waves going along $+\hat{x}$ and $-\hat{x}$. To evaluate b_2 and a_1 in terms of b_1, the incident wave amplitude, we solve Eqs. (3.12) and (3.13) around $\xi \simeq 0$ and match the solutions of WKB solutions for large ξ.

Since the scale length of variation of G is very long, we may replace $G(\xi)$ in Eq. (3.12) around $\xi \ll 1$ by $G(0)$. Then Eq. (3.12) yields

$$E_x = \frac{k_z\omega}{c}\left(\frac{c}{v_{\rm th}}\right)^{2/3}\lambda_{\rm em}^2 G(0) W\left(\frac{c^{2/3}}{v_{\rm th}^{2/3}}\xi\right) \tag{3.16}$$

where $W(\eta)$ is a solution of the inhomogeneous equation $W'' - \eta W = 1$,

$$W(\eta) = -i\int_0^\infty e^{-i(t\eta + t^3/3)}dt \ . \tag{3.17}$$

For $(-\eta) \gg 1$,

$$W(\eta) \simeq (-\eta)^{-1/4}\pi^{1/2}e^{-i\pi/4 + i2/3(-\eta)^{3/2}} - \frac{1}{\eta} \ . \tag{3.18}$$

Matching solution (3.16) to (3.14) asymptotically we get

$$a_1 = \frac{k_z\omega}{c}\left(\frac{c}{v_{\rm th}}\right)^{3/2}\lambda_{\rm em}^2 \pi^{1/2}e^{-i\pi/4}G(0) \ . \tag{3.19}$$

Figure 3.2 illustrates the mode structure of the field component E_x.

Regarding Eq. (3.13), one must notice that its right hand side, for $\xi > v_{\rm th}^2/c^2$, where Eq. (3.14) could be used for E_x, is proportional to the plasma wave amplitude a_1 and goes as $\exp\left[(ic/v_{\rm th})(2/3)(-\xi)^{3/2}\right]$ which is too rapid for the transverse wave. Consequently, the contribution of the right hand side, beyond $\xi > v_{\rm th}^2/c^2$, to G is very small. Let $G_1(\xi)$ and $G_2(\xi)$ be the solutions of the homogeneous equation (3.13), i.e., when its right hand side is put to zero. Then the solution of Eq. (3.13) can be written as

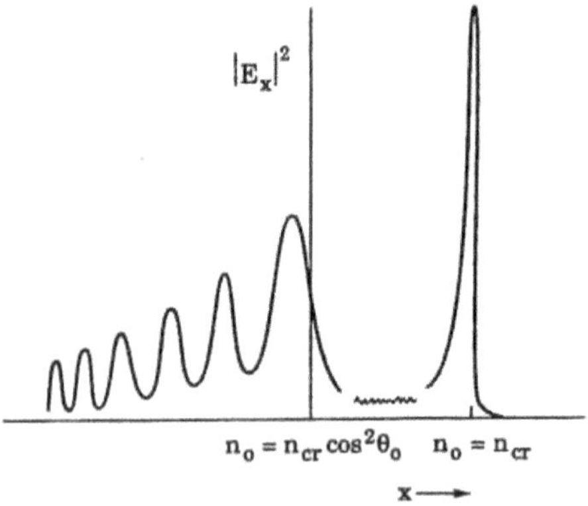

Fig. 3.2. The mode structure of the field component E_x near the turning point ($n_0 = n_{cr}\cos^2\theta_0$) and the critical layer ($n_0 = n_{cr}$).

$$G(\xi) = C_1G_1(\xi) - qG_1(\xi)\int_{-\infty}^{\xi} \frac{G_2(\xi')}{\xi'}\left[G(\xi') + \frac{\xi' ck_z}{qw}E_x(\xi')\right]d\xi'$$

$$- qG_2(\xi)\int_{\xi}^{\infty} \frac{G_1(\xi')}{\xi'}\left[G(\xi') + \xi'\frac{ck_z}{wq}E_x(\xi')\right]d\xi' , \qquad (3.20)$$

where $q = k_z^2\lambda_{\rm em}^2$ and one must remember that $G(\xi)$ is a function of q. Since the contribution to the integrals arises only from the vicinity of $\xi' \lesssim v_{\rm th}^2/c^2$, one may replace $G_1(\xi')$ and $G_2(\xi')$ inside them by $G_1(0)$, $G_2(0)$. Further, from Eqs. (3.12) and (3.16)

$$G(\xi') + \xi'\frac{ck_z}{q\omega}E_x(\xi') = G(0)\frac{d^2W}{d\eta'^2} .$$

Hence

$$G(\xi) = C_1G_1(\xi) - qG_1(\xi)G_2(0)G(0)\int_{-\infty}^{\eta} \frac{1}{\eta'}\frac{d^2W}{d\eta'^2}d\eta'$$

$$- qG_2(\xi)G_1(0)G(0)\int_{\eta}^{\infty} \frac{1}{\eta'}\frac{d^2W}{d\eta'^2}d\eta' \qquad (3.21)$$

and

$$G(0) = C_1G_1(0) - qG_1(0)G_2(0)G(0)\int_{-\infty}^{\infty} \frac{1}{\eta'}\frac{d^2W}{d\eta'^2}d\eta' \qquad (3.22)$$

where $\eta = \xi c^{2/3}/v_{\rm th}^{2/3}$ and the principle value of the integral is to be considered;

$$\int_{-\infty}^{\infty} \frac{1}{\eta'} \frac{d^2W}{d\eta'} d\eta' = i \int_0^{\infty} dt \int_{-\infty}^{\infty} \frac{t^2}{\eta'} \exp\left[-i\left(\eta' t + \frac{t^3}{3}\right)\right] d\eta'$$

$$= \int_0^{\infty} dt\ t^2 e^{-it^3/3} \int_{-\infty}^{\infty} \frac{\sin\alpha}{\alpha} d\alpha$$

$$= -i\pi \ , \tag{3.23}$$

$$G(0) = C_1 \frac{G_1(0)}{1 - i\pi q G_1(0) G_2(0)} \ . \tag{3.24}$$

$G_1(0)$ and $G_2(0)$ are functions of q

$$G_1(0) = A_i(q) \left[-\frac{2A_i(q)}{\pi A_i'(q)}\right]^{1/2}$$

$$G_2(0) = B_i(q) \left[\frac{2B_i(q)}{\pi B_i'(q)}\right]^{1/2}$$

where A_i and B_i are the Airy functions and prime denotes their derivatives with respect to q. For $(-\xi) \gg 1$, Eq. (3.21) gives

$$G(\xi) = C_1 G_1(\xi) + i\pi q G_1(0) G(0) G_2(\xi) \ . \tag{3.25}$$

In this limit

$$G_1(\xi) = (-\xi - q)^{1/4} \sin\left[\frac{3}{2}(-\xi - q)^{3/2} + \delta_0\right]$$

$$G_2(\xi) = (-\xi - q)^{-1/4} \cos\left[\frac{2}{3}(-\xi - q)^{3/2} + \delta_0\right] \ ;$$

hence, $G(\xi)$ can be cast in the form (3.15) giving

$$b_{1,2} = \frac{ic_1}{2} \left\{ \frac{1 \pm \pi q G_1^2(0) - i\pi q G_1(0) G_2(0)}{1 - i\pi q G_1(0) G_2(0)} \right\} e^{\mp i\delta_0} \tag{3.26}$$

with δ_0 as a real constant. The absorption coefficient can now be obtained as

$$A = 1 - \left|\frac{b_2}{b_1}\right|^2 \tag{3.27}$$

which depends only on a single parameter $q \equiv (\omega L_n/c)^{2/3} \sin^2\theta_o$. Figure 3.3 demonstrates the variation of A with q. A maximizes to ~ 0.5 at $q \sim 0.3$. Such behavior can be understood as follows. For mode conversion one requries (i) a component of wave field along the density gradient

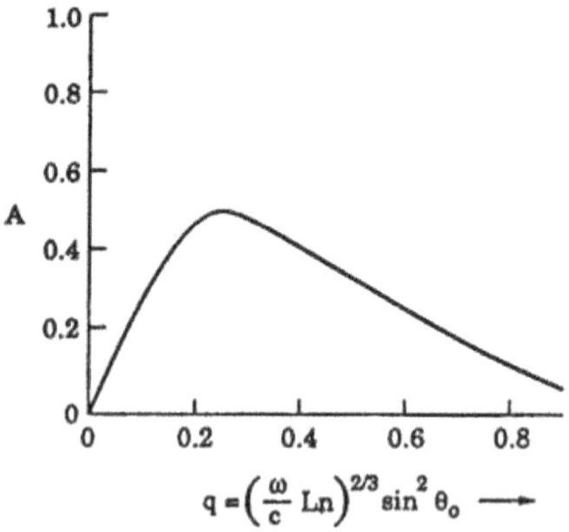

Fig. 3.3. Absorption coefficient as a function of parameter q.

($E_x \neq 0$) and (ii) the separation $|x_T| = L_N \sin^2\theta_0$ between the critical layer and turning point to be small. At normal incidence $\theta_0 = 0$, $q = 0$, the electric vector $\mathbf{E}\|\hat{z}$, hence no density oscillations and no mode conversion. At large angles of incidence the separation between th critical layer and turning point is very large, hence the electromagnetic wave field cannot tunnel to the critical layer and no mode conversion can take palce. The maximum mode conversion (hence maximum absorption) should occur when $|x_T| \sim \lambda_{\rm em} \equiv (c^2 L_n/\omega^2)^{1/3}$, i.e., $(L_N\omega/c)^{2/3}\sin^2\theta_o \sim 1$ which may be compared to the value of $q \sim 0.3$ obtained through rigorous calculation.

Resonance absorption constitutes one of the most important absorption mechanisms in a laser produced plasma.[3] In the presence of a self-generated magnetic field, transverse to the density gradient, the electromagnetic wave propagating in the x-mode possesses finite E_x even at normal incidence and strong absorption may occur near the upper hybrid resonance layer.[4-5] Woods *et al.*[6] and Cairns and Lashmore-Davies[7] have developed an elegant formalism of mode conversion of electromagnetic waves in an inhomogeneous magnetized plasma. One is also referred to the classical paper[8] and the book by Stix.[9]

References

1. V. L. Ginzburg, *Propagation of Electromagnetic Waves in Plasma* (Pergamon Press, 1960).
2. A. D. Piliya, *Sov. Phys. Tech. Phys.* **11**, 609 (1966).
3. W. Kruer, *The Physics of Laser Plasma Interactions* (Addison-Wesley, 1987).
4. C. Grebogi, C. S. Liu, and V. K. Tripathi, *Phys. Rev. Lett.* **39**, 338 (1977).

5. R. W. White and F. F. Chen, *Plasma Phys.* **16**, 565 (1974).
6. A. M. Woods, R. A. Cairns, and C. N. Lashmore-Davies, *Phys. Fluids* **29**, 3719 (1986).
7. R. A. Cairns and C. N. Lashmore-Davies, *Phys. Fluids* **26**, 1268 (1983).
8. T. H. Stix, *Phys. Rev. Lett.* **15**, 878 (1965).
9. T. H. Stix, *The Theory of Plasma Waves* (McGraw Hill, 1962).

CHAPTER 4

PLASMA WAVE EXCITATION BY TWO LASER BEATING AND PARTICLE ACCELERATION

There exists a great deal of interest in the excitation of collective modes in a plasma by beating two laser radiations, particularly plasma wave for its use in the collective acceleration of charged particles to very high energies[1-11] ($\gtrsim$ GeV). Because the laser frequency is typically much higher than the plasma frequency, one can excite a large amplitude Langmuir wave by beating two collinear laser beams having frequency difference $\Delta\omega \simeq \omega_p$, the plasma frequency. The Langmuir wave of phase velocity near the velocity of light can trap "nearly resonant" charged particles around potential energy minima and accelerate them to a velocity equal to the phase velocity of the Langmuir wave. We analyze this problem in two parts: (I) Resonant excitation of a Langmuir wave and (II) Acceleration of charged particles by the Langmuir wave.

4.1. Excitation of a Langmuir Wave

Consider a uniform unmagnetized plasma of density n_0 and electron temperature T_e. Two collinear laser beams of large amplitude propagate through it[1] (cf. Fig. 4.1):

$$\begin{aligned} \mathbf{E}_1 &= \hat{x}A_1 e^{-i(\omega_1 t - k_1 z)}, \\ \mathbf{E}_2 &= \hat{x}A_2 e^{-i(\omega_2 t - k_2 z)}, \end{aligned} \tag{4.1}$$

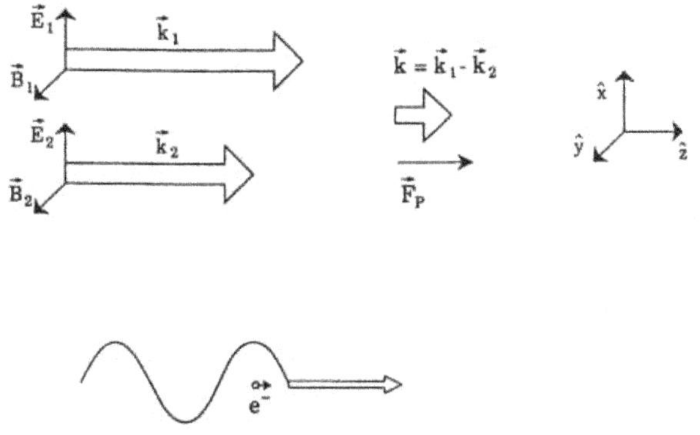

Fig. 4.1. Collinear laser beams, beat ponderomotive force and an electron in the presence of a Langmuir wave.

where $\omega_1 - \omega_2 \simeq \omega_p$, $\omega_1 \gg \omega_p$, $\omega_2 \gg \omega_p$. They produce oscillatory electron velocities $\mathbf{v}_j = e\mathbf{E}_j/mi\omega_j$ $(j = 1, 2)$ and exert a ponderomotive force $\mathbf{F}_p = -\frac{e}{2c} \cdot (\mathbf{v}_1 \times \mathbf{B}_2^* + \mathbf{v}_2^* \times \mathbf{B}_1) = \hat{z}ike\phi_p$ on them where,

$$\phi_p = -\frac{m}{2e}\mathbf{v}_1 \cdot \mathbf{v}_2^* = \Phi_p e^{-i(\omega t - kz)} , \tag{4.2}$$

$\Phi_p = eA_1A_2/2m\omega_1\omega_2$, $\omega = \omega_1 - \omega_2$, $k = k_1 - k_2$, $\mathbf{B}_j = c\mathbf{k}_j \times \mathbf{E}_j/\omega_j$ and $-e$ and m are the electronic charge and mass. In writing Φ_p we have used the fact that all physical observables $\mathbf{E}_j, \mathbf{B}_j, \mathbf{v}_j$ etc. are actually the real parts of their complex representation and that $\mathrm{Re}\,A\,\mathrm{Re}\,B \equiv \frac{1}{2}\,\mathrm{Re}(AB + AB^*)$. One may notice that the phase velocity of the ponderomotive force equals the group velocity of the laser beams, $\omega/k = c(1 - \omega_p^2/\omega_1^2)^{1/2} \approx c$. It drives a Langmuir wave whose electrostatic potential can be taken as

$$\phi = \Phi e^{-i(\omega t - kz)} . \tag{4.3}$$

Φ, in general, is complex with slow space-time dependence. The response of electrons to ϕ and ϕ_p can be obtained treating Φ as a constant and later replacing ω and k by time-space operators. The density perturbation can be derived in a way similar to that used in Sec. 2.7 with $\nabla\phi_p$ as additional force:

$$n = \frac{k^2}{4\pi e}\chi_e(\phi + \phi_p) \tag{4.4}$$

where $\chi_e = -\left[\omega_p^2 + 3k^2 v_{th}^2/2\right]/\omega^2$ is the electron susceptibility. Using Eq. (4.4) in the Poisson's equation $\nabla^2\phi = 4\pi en$, we obtain

$$\varepsilon\Phi = -\chi_e \Phi_p \tag{4.5}$$

where $\varepsilon = 1 + \chi_e$. For resonant excitation $\varepsilon \approx 0$,

$$\omega^2 = \omega_p^2 + \frac{3}{2}k^2 v_{th}^2 \tag{4.6}$$

one must notice that since $k \sim \omega/c$, $\omega \simeq \omega_p$, a small nonuniformity in electron density would cause a strong modification in k, i.e., a strong wave number mismatch, limiting the region of resonant excitation. We solve Eq. (4.5) in different cases.

(A) *Temporal growth in a homogeneous plasma*

Replacing ω by $\omega + i\partial/\partial t$ and $\varepsilon(k,\omega) = i(\partial\varepsilon/\partial\omega)(\partial/\partial t)$, Eq. (4.5) can be written as

$$\frac{\partial}{\partial t}\Phi = -i\frac{\omega}{2}\Phi_p\ . \tag{4.7}$$

Expressing $\Phi = A(t)e^{-i\psi(t)}$ and separating real and imaginary parts, Eq. (4.7) yields

$$\dot{A}\cos\psi - A\dot{\psi}\sin\psi = 0,$$

$$A\sin\psi + A\dot{\psi}\cos\psi = \frac{\omega}{2}\Phi_p$$

i.e.,

$$\dot{A} = \frac{\omega}{2}\Phi_p\sin\psi,$$

$$A\dot{\psi} = \frac{\omega}{2}\Phi_p\cos\psi\ . \tag{4.8}$$

One must remember that ψ is the phase of the Langmuir wave with respect to the ponderomotive wave (also called the beat wave). Since $\dot{\psi}$ is positive for $-\pi/2 < \psi < \pi/2$ and negative for $\pi/2 < \psi < 3\pi/2$ the Langmuir wave rapidly phase locks with the beat wave with stationary phase $\psi = \pi/2$. The stationary phase is reached when $\cos\psi = 0$ and $\dot{A} > 0$. The Langmuir wave grows with time,

$$A(t) = A(0) + \frac{\omega}{2}\Phi_p t\ . \tag{4.9}$$

As the amplitude increases, the relativistic mass correction becomes important, introducing a frequency mismatch and leading to phase unlocking and saturation of the amplitude. In this case the equation of motion can be written as[1]

$$\frac{\partial}{\partial t}(\gamma v_z) + v_z \frac{\partial}{\partial z}(\gamma v_z) = \frac{e}{m}\frac{\partial}{\partial z}(\phi + \phi_p) \tag{4.10}$$

where $\gamma \cong 1 + v_z^2/2c^2$. To the lowest order

$$v_z = -\frac{ek}{m\omega}(\phi + \phi_p)\ . \tag{4.11}$$

One must realize that $\phi \gg \phi_p$ for resonant excitation (cf. Eq. (4.5)). Further,

$$v_z^2 = \frac{1}{2}\operatorname{Re}[v_z v_z^* + v_z v_z]$$

$$v_z^2 v_z = \frac{1}{4}\operatorname{Re}[v_z v_z^* v_z + v_z v_z v_z + v_z^* v_z v_z + v_z v_z v_z^*]\ . \tag{4.12}$$

Employing Eq. (4.12) in Eq. (4.10) v_z can be obtained to the next order,

$$\begin{aligned} v_z &= -\frac{ek}{m\omega}(\phi + \phi_p) - \frac{1}{2c^2}\frac{3}{4}v_z^2 v_z^* \\ &\simeq -\frac{ek}{m\omega}(\phi + \phi_p) + \frac{3}{8c^2}\frac{e^3k^3}{m^3\omega^3}\phi^2\phi^*\ . \end{aligned} \tag{4.13}$$

The electron density can be written as

$$\begin{aligned} n &= n_0\frac{kv_z}{\omega} \\ &= -n_0\frac{k^2e}{m\omega^2}(\phi + \phi_p) + \frac{3n_0e^3k^4}{8c^2m^3\omega^4}\phi^2\phi^*\ . \end{aligned} \tag{4.14}$$

Here we have neglected the nonlinearity arising through the equation of continuity.[2] Using n in the Poisson's equation, we get

$$\varepsilon\Phi = -\chi_e\Phi_p - \frac{3}{8}\frac{e^2k^2\Phi^2\Phi^*}{m^2\omega^2c^2} \tag{4.15}$$

which on expressing $\omega \rightarrow \omega + i\partial/\partial t$ leads to

$$\frac{\partial\Phi}{\partial t} = -i\frac{\omega}{2}\Phi_p + i\frac{3\omega}{16}\frac{e^2k^2\Phi^2\Phi^*}{m^2\omega^2c^2} \tag{4.16}$$

Expressing $\Phi = A(t)e^{-i\psi(t)}$, one obtains from Eq. (4.16)

$$\dot{A} = \frac{\omega}{2}\Phi_{\rm p} \sin\psi \ ,$$

$$A\dot{\psi} = \frac{\omega}{2}\Phi_{\rm p} \cos\psi - \alpha A^3 \tag{4.17}$$

where $\alpha = 3e^2k^2/16m^2\omega c^2$. We attempt a solution of Eqs. (4.17) around $\Psi = \pi/2$; $\Psi = \pi/2 - f$; where $f \ll \pi/2$. Dividing the second equation by the first we get

$$A\frac{df}{dA} \cong -f + \frac{2\alpha}{\omega\Phi_{\rm p}}A^3 \ . \tag{4.18}$$

Introducing $A = e^x$, Eq. (4.18) takes the form

$$\frac{df}{dx} = -f + \frac{2\alpha}{\omega\Phi_{\rm p}}e^{3x} \ , \tag{4.19}$$

giving

$$f = \frac{1}{2}\frac{\alpha}{\omega\Phi_{\rm p}}A^3$$

for the case $A(0) \ll \Phi_{\rm p}$. Hence,

$$\begin{aligned}\dot{A} &= \frac{\omega}{4}\Phi_{\rm p}(2 - f^2) \\ &= \frac{\omega}{2}\Phi_{\rm p}\left[1 - \frac{1}{32}\left(\frac{2\alpha A^3}{\omega\Phi_{\rm p}}\right)^2\right] \ .\end{aligned} \tag{4.19a}$$

The wave growth stops when

$$A^3 = \frac{\omega\Phi_{\rm p}}{\alpha}2\sqrt{2} \tag{4.20}$$

i.e., when $|v_z|/c = \{32\sqrt{2}v_1v_2^*/c^2\}^{1/3}$. The oscillatory electron velocity due to the Langmuir wave is larger than v_1, v_2 which justifies our neglect of relativistic mass corrections due to v_1, v_2. Further, since $k \sim \omega/c$, $k|v_z|/\omega \ll 1$ wave breaking is unimportant. A more rigorous approach, using a Lagrangian variable, can be found in Ref. 1.

(B) *Excitation in an inhomogeneous plasma*

Since the k-vector of the Langmuir wave is sensitive to plasma density, $k^2 = 2(\omega^2 - \omega_{\rm p}^2)/3v_{\rm th}^2$, the region of resonant excitation is limited even in a gentle density gradient. Consider a linear density profile

$$\omega_{\rm p}^2 = (\omega_1 - \omega_2)^2\left(1 + \frac{z}{L_n}\right) \ . \tag{4.21}$$

The phase matching condition, $k = k_1 - k_2 \cong (\omega_1 - \omega_2)/c$ is satisfied at $z = -(3v_{\rm th}^2/2c^2)L_n$. At lower densities k goes as $\frac{\omega_p}{v_{th}}(-z/L_n)^{1/2}$ whereas it vanishes at $z = 0$. Thus the nonlinear excitation is spatially localized. The Langmuir wave should reach a steady state when the rate of energy convection fro the interaction region equals the late of energy supplied by the laser. To investigate the steady state behavior we solve the spatial problem, including thermal effects. Since the phase matching and turning points are close to each other one must solve the wave equation without WKB approximation. In the presence of the ponderomotive field $\mathbf{E}_{\rm p} = -\hat{z}i(\omega/c)\phi_{\rm p}$ and the field $\mathbf{E} = -(\partial\phi/\partial z)\hat{z}$ of the Langmuir wave the electron oscillatory velocity, on solving the equations of continuity and motion, turns out to be

$$v_z = \frac{eE_z}{mi\omega} - \frac{3}{2}\frac{v_{\rm th}^2}{\omega^2}\frac{e}{mi\omega}\frac{\partial^2 E_z}{\partial z^2} + \frac{eE_{\rm p}}{mi\omega} \tag{4.22}$$

Using v_z in the Fourth Maxwell equation, viz., $\partial E_z/\partial t - 4\pi n_0 e v_z = 0$ we obtain

$$\frac{d^2E_z}{dz^2} + \frac{2}{3}\frac{\omega^2 - \omega_{\rm p}^2}{v_{\rm th}^2}E_z = -\frac{2}{3}\frac{\omega^2}{v_{\rm th}^2}E_{\rm p}e^{i\frac{\omega}{c}z}$$

or

$$\frac{d^2E_z}{d\xi^2} - \xi E_z = -\frac{2}{3}\frac{L_n}{\lambda_{\rm es}}E_{\rm p}e^{i\frac{\omega}{c}\lambda_{es}\xi} \tag{4.23}$$

where $\xi = z/\lambda_{\rm es}$, $\lambda_{\rm es} = (3v_{\rm th}^2 L_n/2\omega^2)^{1/3}$, $\phi = \Phi e^{i\omega t}$. Equation (4.23) has a general solution, well behaved at $\xi \to +\infty$,

$$E_z = C_1 A_i(\xi) + \frac{2}{3}\frac{L_n}{\lambda_{\rm es}}E_{\rm p}\pi$$

$$\times\left[A_i(\xi)\int_{-\infty}^{\xi} B_i(\xi)e^{i\frac{\omega}{c}\lambda_{es}\xi}d\xi - B_i(\xi)\int_{\infty}^{\xi} A_i(\xi)e^{i\frac{\omega}{c}\lambda_{es}\xi}d\xi\right]. \tag{4.24}$$

At $\xi \to -\infty$ we must have only an outgoing Langmuir wave. Since,

$$A_i(\xi \to -\infty) \simeq \pi^{-1/2}(-\xi)^{-1/4}\sin\left[\frac{2}{3}(-\xi)^{3/2} + \frac{\pi}{4}\right],$$

$$B_i(\xi \to -\infty) \simeq \pi^{-1/2}(-\xi)^{-1/4}\cos\left[\frac{2}{3}(-\xi)^{3/2} + \frac{\pi}{4}\right],$$

this condition, along with the realization that ϕ has a time-dependence $\sim e^{-i\omega t}$, demands

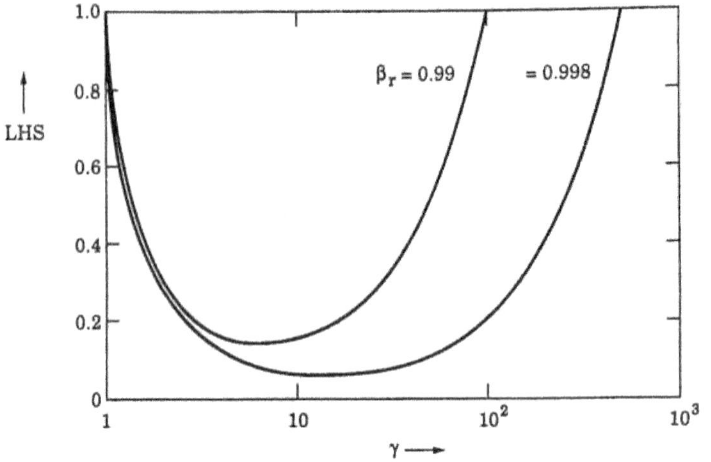

Fig. 4.2. Variation of LHS $\equiv \gamma - \beta_r(\gamma^2 - 1)^{1/2}$ as a function of γ.

$$C_1 = i\pi \frac{2}{3}\frac{L_n}{\lambda_{es}} E_p \int_{-\infty}^{\infty} A_i(\xi) e^{i\frac{\omega}{c}\lambda_{es}\xi} d\xi$$

$$= i\pi \frac{2}{3}\frac{L_n}{\lambda_{es}} E_p e^{-\frac{i}{3}(\frac{\omega}{c}\lambda_{es})^3} . \qquad (4.25)$$

Thus the amplitude of the outward-going Langmuir wave is

$$E_z = \pi \frac{2}{3}\frac{L_n}{\lambda_{es}}\frac{\omega}{c}\Phi_p \pi^{-1/2}(-\xi)^{-1/4} \qquad (4.26)$$

which yields the power flux density

$$P_{out} = \omega \frac{\partial \varepsilon}{\partial k}\frac{E_z E_z^*}{16\pi} = \left(\omega^3 \frac{L_n}{16c^2}\right)\Phi_p^2$$

4.2. Electron Acceleration in a Langmuir Wave

Now we investigate single particle acceleration in a large amplitude Langmuir wave

$$\mathbf{E} = -\hat{z}E_0 \sin(\omega_p t - k_p z + \psi_0) , \qquad (4.27)$$

following Bobin.[7] In Eq. (4.27) ψ_0 is initial phase. We ignore x-y particle motions. The momentum equation for an electron $dp/dt = -eE_z$ can be cast into the following form

$$\frac{d\gamma}{dt} = \frac{p}{\gamma m^2c^2}\frac{dp}{dt} = -\frac{peE_z}{\gamma m^2c^2} , \tag{4.28}$$

where $\gamma = \sqrt{1+p^2/m^2c^2}$ is the relativistic gamma factor and m is the rest mass. Introducing a similarity variable

$$\xi = k_\text{p}z - \omega_\text{p}t - \beta_0 \tag{4.29}$$

one gets the differential system

$$\frac{d\gamma}{dt} = -\frac{eE_0}{mc}\left(1-\frac{1}{\gamma^2}\right)^{1/2}\sin\xi \tag{4.30}$$

$$\frac{d\xi}{dt} = -ck_0\left(1-\frac{1}{\gamma^2}\right)^{1/2} - \omega_\text{p} . \tag{4.31}$$

ξ is the phase of the wave as seen by the electron and $(\gamma-1)mc^2$ is the kinetic energy of the electron. Dividing $d\gamma/dt$ by $d\xi/dt$ and integrating the resulting equation, we obtain

$$\gamma - \beta_r\left(\gamma^2-1\right)^{1/2} = A\cos\xi + C \tag{4.32}$$

where $\beta_r = \omega_\text{p}/k_\text{p}c$, $A = eE_0/mc^2k_\text{p}$, $C = \gamma_0 - \beta_r\left(\gamma_0^2-1\right)^{1/2} - A\cos\psi_0$ is the constant of integration and $(\gamma_0 - 1)mc^2$ is the electron energy at the initial phase ξ. Equation (4.32) gives the γ, ξ phase space behavior. One must recall that $\omega_\text{p}/k_\text{p}$, the phase velocity of the Langmuir wave, is less than c, hence the second term on the left hand side (LHS) is less than the first and the LHS is a positive definite. For $C < A$ only those values of ξ are accessible for which RHS is positive. Such particles are trapped particles. Further, for one value of LHS (or RHS) there are two values of γ (cf. Fig. 4.2), hence, for one value of ξ, r has two values. These values coalesce when

$$\frac{d}{d\gamma_0}\ (\text{LHS}) = 0 ,$$

i.e.,

$$\gamma = \frac{1}{\sqrt{1-\beta_r^2}}, \qquad \cos\xi - \frac{\sqrt{1-\beta_r^2}-c}{A} . \tag{4.33}$$

For $C > A + \sqrt{1-\beta_r^2}$ all values of ξ are accessible, representing untrapped particles. Figure 4.2 displays phase space trajectories of trapped and untrapped particles. One must note that trapped particle trajectories have

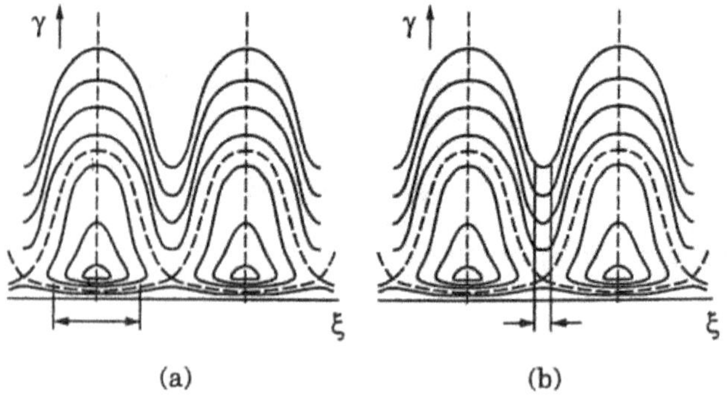

Fig. 4.3. γ, ξ phase plot: (a) acceleration of trapped electrons, (b) acceleration of passing electrons. [After Bobin[7]].

large γ width because as seen in Fig. 4.2, the two values of γ, corresponding to a given value of LHS or of ξ, are far apart. Figure 4.3 shows that the following can be accelerated to high energies (high γ): (i) electrons with trapped trajectories and initial low energy and (ii) electrons with sharply defined initial phase on passing trajectories associated with velocities greater than the phase velocity of the wave.

Acceleration energy and length

Using the value of C in terms of initial energy and phase as γ_0 and ξ_0, Eq. (4.32) can be written as

$$\gamma - \gamma_0 - \beta_r\left[\left(\gamma^2 - 1\right)^{1/2} - \left(\gamma_0^2 - 1\right)^{1/2}\right] = A(\cos\xi - \cos\xi_0)\,. \qquad (4.34)$$

The largest value of the RHS is $2A$. If we assume $\gamma, \gamma_0 \gg 1$ (as is true for passing trajectories), then Eq. (4.34) gives the maximum energy W_A that an electron may acquire

$$\begin{aligned}\Delta\gamma \equiv \gamma - \gamma_0 &= \frac{2eE_0}{mc\omega_p}\frac{\beta_r}{1-\beta_r}\,,\\ W_A &= mc^2\Delta\gamma\\ &\cong \frac{4eE_0c}{\omega_p}\gamma_r^2\,,\end{aligned} \qquad (4.35)$$

where $\gamma_r^2 = (1-\beta_r^2)^{-1}$. In Ref. 3 the maximum value E_0 was assumed to be limited by wave breaking, corresponding to the ratio of perturbed density to unperturbed density i.e., $\lesssim 1/2$,

$$\frac{n}{n_0^0} \sim \frac{ek_p E_0}{m\omega_p^2} \simeq \frac{1}{2}$$

i.e.,

$$E_0 \sim \frac{m\omega_p^2}{2ek_p} \sim \frac{mc\omega_p}{2e} . \tag{4.36}$$

Hence,

$$W_{A\,\mathrm{max}} \simeq 2m_0c^2\gamma_r^2 , \tag{4.37}$$

In the moving frame, electrons are accelerated over almost half a wavelength. Beyond that distance they are decelerated. The acceleration length L_A in the laboratory frame is given by

$$W_A = eE_0L_A$$

or

$$L_A = 2\gamma_r^2 c/\omega_p . \tag{4.38}$$

These are the upperbounds in the energy gain by an electron in a Langmuir wave. However, the amplitude is usually limited by the relativistic effects to a much lower level than that given by $n/n_0 = 1/2$ as discussed in Sec. 4.1A, Eq. (4.20) and W_A could be much smaller.

Surfatron

The main limitation in the above acceleration process results from wave-particle dephasing. For given plasma and laser conditions, it puts an upper bound on the energy a particle can acquire. One may overcome this limitation by superimposing on the plasma wave a uniform magnetic field in a direction perpendicular to the wave vector. The magnetic field deflects particles across the wavefront, thus preventing them from outrunning the wave. One may view the process as follows. Consider a trapped electron moving with velocity $v_z\hat{z} \simeq \omega_p/k_z\hat{z}$ i.e., the phase velocity of the wave. As there exists a static magnetic field $B_s\hat{y}$, the electron experiences a $ev_zB_s/c\hat{x}$ force and acquires v_x velocity in the x direction. v_x and B_s give rise to a retarding force $F_z = -ev_xB_s/c$ opposite to the z direction. If the electric force $-eE_z$ on the electron due to the wave is sufficient to overcome the retarding force, the electron would remain trapped and get accelerated in

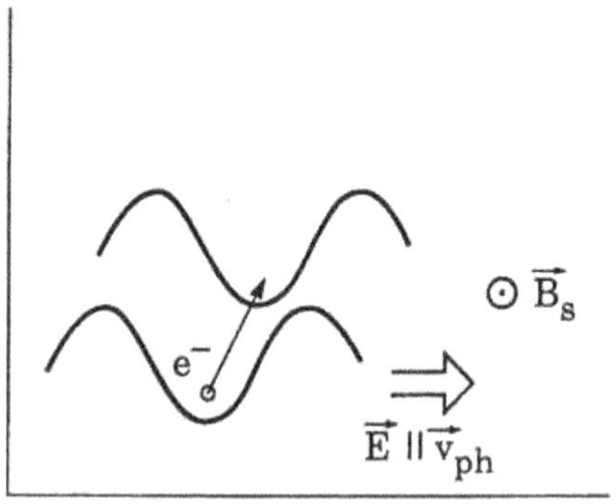

Fig. 4.4. Electron trapped in the potential well of a Langmuir wave is acceleration across the wavefront by the $\mathbf{v}_{ph} \times \mathbf{B}_s$ force.

the x direction without bound. One might remember that since $\mathbf{v} \times \mathbf{B}$ force does not do any work on the electron, all the energy it gains comes from the wave, which has to continuously exert a forward force on the electron to counter the magnetic retarding force. Katsoules and Dawson[4] have shown that the particles may be accelerated to arbitrarily high energy as they ride across the wavefront, like surfers cutting across the face of an ocean wave (Fig. 4.4).

Consider the motion of an electron in the electric field of large amplitude Langmuir wave (cf. Eq. (4.27)) and a dc magnetic field $B_s\hat{y}$,

$$\frac{d}{dt}(\gamma v_z) = -\frac{eE_0}{m}\sin(\omega_p t - k_p z + \psi_0) - \omega_c v_x \tag{4.39}$$

$$\frac{d}{dt}(\gamma v_x) = \omega_c v_z \ , \tag{4.40}$$

where $\gamma = (1 - (v_x^2 + v_z^2)/c^2)^{-1/2}$, $\omega_c = eB_s/mc$. We solve this set for a particle trapped in the z-direction. To obtain a criterion for trapping, one may write down the force in the wave frame

$$F_z' = -e\left(E_0 \sin\xi + \gamma_r \frac{v_x'}{c} B_s\right) \ . \tag{4.41}$$

where $\gamma_r = (1 - \omega_p^2/k_p^2c^2)^{-1/2}$ and v_x' is the x velocity in the moving frame. The first term of the force is the trapping term while the second is the gyratory or detrapping term. An initially trapped particle can never detrap if

$$\gamma_r B_s < E_0 \ . \tag{4.42}$$

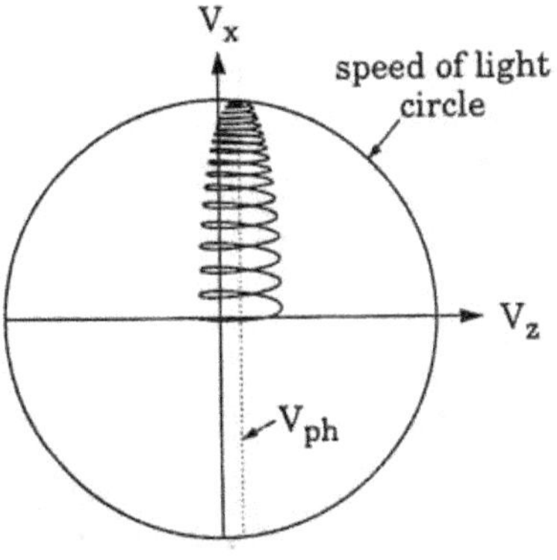

Fig. 4.5. Velocity-space trajectory of a particle in a low-phase-velocity wave (V_{ph} = $0.1c, E_0/B_s = 1.5, \omega/\omega_c = 2$). [After Katsouleas and Dawson.[3]]

Under condition (4.42) we assume that to the zeroth order $v_z \simeq \omega_p/k_p$, then Eq. (4.40) can be integrated to obtain $\gamma v_x = \omega_c \omega_p t/k_p$, or,

$$v_x = \frac{\omega_c \left(\frac{\omega_p}{k_p}\right) t}{\gamma_r \left(1 + \omega_c^2 t^2 \frac{\omega_p^2}{k_p^2 c^2}\right)^{1/2}}, \tag{4.43}$$

for particle motion across the wavefront.

Figure 4.5 shows the velocity space trajectories obtained as a numerical solution of Eqs. (4.39) and (4.40) by Katsouleas and Dawson[4]. The particle's total velocity approaches asymptotically the speed of light circle as predicted by Eq. (4.43). The higher order motion observed in the figure can be represented by the first order linearized version of (4.39), which can be approximately written as

$$\ddot{z}_1 + \frac{\omega_b^2}{\gamma} z_1 \approx -\frac{\gamma_r^2}{\gamma^2} \omega_c^2 \left(\frac{\omega_p}{k_p}\right) t$$

where $\omega_b = \sqrt{eE_0 k/m}$ is the nonrelativistic bounce frequency. This driven oscillator equation describes the bounce motion of a particle in the potential trough of the wave and its shift out of the bottom of the potential well due to the relativistic mass increase and $\mathbf{v} \times \mathbf{B}_s$ force. Carioli *et al.*[5] have recently examined this problem in extensive detail. One is referred to their paper on stochastic and coherent acceleration of charged particles.

References

1. M. N. Resonbluth and C. S. Liu, *Phys. Rev. Lett.* **29**, 701 (1972).
2. Inclusion of the nonlinear term arising from the equation of continuity would have produced an arroneous result as this implies a d.c. current. This rather paradoxial result was discussed by J. M. Dawson, in *From Particles to Plasma*, ed. J. W. Van Dam (Addison-Wesley, Reading, MA 1989) p. 131.
3. T. Tajima and J. M. Dawson, *Phys. Rev. Lett.* **43**, 267 (1979).
4. T. Katsoulea and J. M. Dawson, *Phys. Rev. Lett.* **51**, 392 (1983).
5. S. M. Carioli, A. A. Chernikov, and A. I. Neishtadt, *Phys. Sci.* **40**, 707 (1989).
6. P. Sprangle, E. Esarey, A. Ting, and G. Joyce, *Appl. Phys. Lett.* **53**, 2146 (1988); *ibid* **53**, 1266 (1988).
7. J. L. Bobin, in *Proc. of the ECFA-CAS/CEFN-In-2P3-IRF/CEA-EPS Workshop, Orsay, Italy*, 1987, ed. S. Turnet (CERN, Geneva, 1987), **1**, p. 58.
8. C. E. Clayton, C. Joshi, C. Darrow, and D. Umstadter, *Phys. Rev. Lett.* **54**, 2343 (1985).
9. B. Amini and F. F. Chen, *Phys. Rev. Lett.* **53**, 1441 (1984).
10. W. B. Mori, *IEEE Trans. Plasma Sic.* **PS-15**, 88 (1987).

CHAPTER 5

COHERENT EMISSION OF RADIATION

The generation of high-powered coherent radiation by electron beams has been an important field of research for several decades. Early devices, viz., the Klystron, magnetron and travelling wave tube, employed non-relativistic electron beams producing microwave radiation with high efficiency at centimeter wavelengths ($\lambda \geq 1$ cm). At shorter wavelengths (1 cm $> \lambda >$ 2 mm), gyrotrons, employing mildly relativistic electron beams, have emerged to be very efficient sources of high power *cw* and pulsed radiation. Their operation at shorter wavelengths is limited only by the requirement of a high magnetic field. In recent years free electron lasers and Cerenkov free electron lasers have become very attractive devices at millimeter to infrared wavelengths[1-11] (cf. Fig. 5.1) and are capable of generating even much shorter wavelengths, up to x-rays. The underlying physics of these devices is simple and exciting. We will discuss them at some length.

5.1. Phase Coherence and Bunching

A key element of the emission process is the bunching of the electron beam by the radiation field such that the electrons are decelerated by the field ($\mathbf{J} \cdot \mathbf{E} < 0$) emitting coherent radiation. In a Cerenkov free electron laser a relativistic electron beam propagates through a dielectric loaded waveguide that supports a TM mode, with finite E_z:

$$E_z = A(r)\cos(\omega t - kz)$$

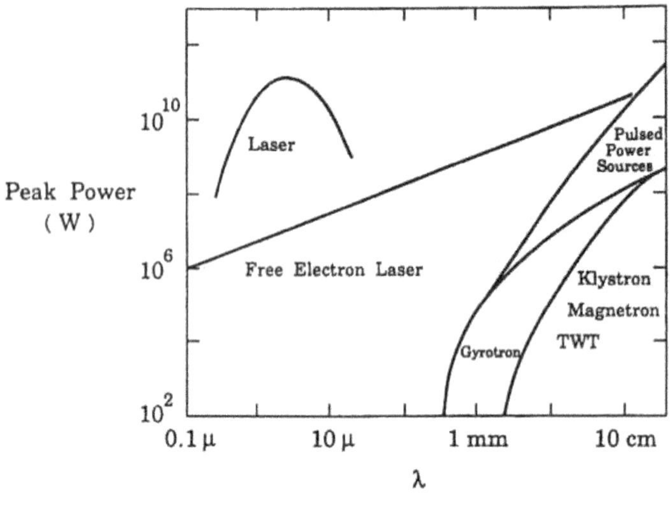

Fig. 5.1. High-powered sources of coherent radiation at submicron to (cm) wavelengths.

having phase velocity $\omega/k < c$. When the velocity of the beam $v_b \simeq \omega/k$ the wave appears to them almost a static field, capable of accelerating or decelerating electrons efficiently. If $v_b = \omega/k$ exactly the field is purely static, accelerating half of the electrons and retarding the other half, resulting in no net loss or gain of energy from the particles. The situation is different when v_b is slightly greater than ω/k. One may conveniently visualize the interaction in a frame moving with velocity $\hat{z}\omega/k$, where $E_z = A(r)\cos(k'z')$ and the particle velocity $v_b' = (v_b - \omega/k)/(1 - v_0\omega/k_z c^2) > 0$. There are two kinds of regions (cf. Fig. 5.2: I) the accelerating zones (where $-eE_z > 0$) and II) the decelerating zones (where $-eE_z < 0$). Initially electrons are uniformly distributed at all z'. However, the ones in zone I are accelerated and quickly move over to zone II, whereas those in zone II are retarded, spending more time there. Thus there is a net build-up of electrons in the retarding zone, resulting in a net transfer of energy from them to the wave.

Quantum mechanically the excitation of radiation can be viewed as a process in which an electron with initial energy $\varepsilon = mc^2\gamma_0$ and momentum $\mathbf{p} = m\gamma_0\mathbf{v}_b$ (m being the rest mass and γ_0 the relativistic gamma factor) goes over to $\varepsilon - \hbar\omega$ and $\mathbf{p} - \hbar\mathbf{k}$, emitting a photon of energy $\hbar\omega$ and momentum $\hbar\mathbf{k}$. Since

$$\varepsilon - \hbar\omega = mc^2\left[1 + \frac{(\mathbf{p} - \hbar\mathbf{k})^2}{m^2c^2}\right]^{1/2},$$

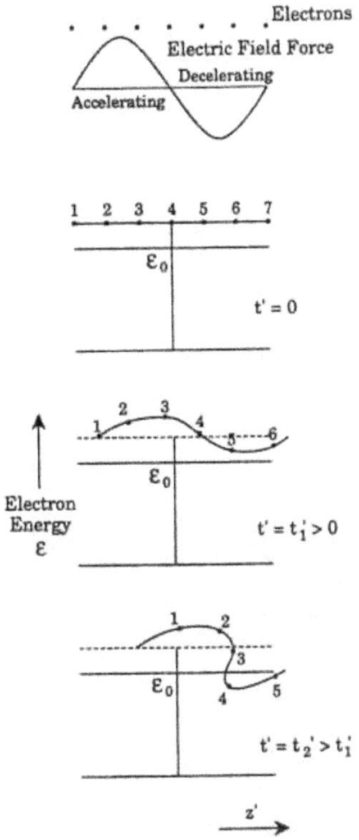

Fig. 5.2. Energy and axial position of electrons at different instants of time. Initially all electrons have the same energy $\varepsilon_{\rm in} > \varepsilon_0$ where ε_0 is the energy of a resonant electron moving with velocity ω/k with respect to the lab frame.

$\varepsilon = mc^2\left(1 + p^2/m^2c^2\right)^{1/2}$ we obtain (for $\hbar k \ll p$)

$$\omega = \mathbf{k} \cdot \mathbf{v}_{\rm b} \,, \tag{5.1}$$

the same condition as for wave-particle phase synchronism. In an unmagnetized plasma $k < \omega/c$, hence, this condition cannot be satisfied, and one needs a decelerating medium. Equation (5.1) must be viewed along with the electromagnetic wave dispersion relation $\omega(k)$ in the medium. In a partially dielectric loaded waveguide, for example, the dispersion relation for a TM mode is illustrated in Fig. 5.3. The phase velocity of the wave is ∞ at the waveguide cutoff and asymptotically approaches $c/\sqrt{\varepsilon}$ (ε being the permittivity of the dielectric fillings) at high frequencies. At some intermediate frequency, i.e., at the intersection of the dispersion curve and the

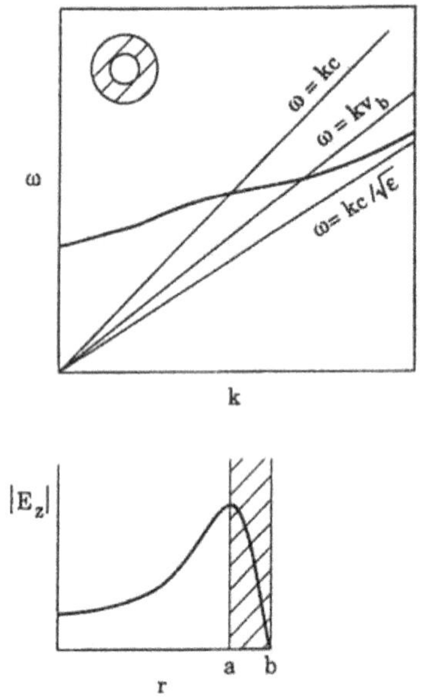

Fig. 5.3. Dispersion relation and mode structure of a TM mode in a dielectric lined cylindrical waveguide. The field of a slow mode ($\omega < kc$) peaks inside the dielectric.

$\omega = kv_b$ line (called the beam mode), the phase velocity equals the beam velocity. This is the operating frequency of the device. Since the waveguide supports several TM modes, in principle, many discrete frequencies can be generated. However, fields of all the modes having phase velocity less than c peak inside the dielectric, falling off rapidly away from it; the higher the mode number, the more severe the decay. Hence, only the field of the fundamental mode is primarily seen by the beam, leading to single mode excitation.

In a free electron laser a relativistic electron beam propagates through a wiggler magnetic field (produced by a suitable arrangement of magnets, or by a pair of coils, cf. Fig. 5.4): $\mathbf{B} = B_0(\hat{x}\cos|k_0|z + \hat{y}\sin|k_0|z)$. When an electromagnetic signal $(\omega, k\hat{z})$ exists in the interaction region the electrons experience a $\mathbf{v}\times\mathbf{B}$ ponderomotive force along z-axis (due to the wiggler and the signal): $F = F_0\cos(\omega t - (k+|k_0|)z)$ having phase velocity $\omega/(k+|k_0|)$. A near-phase synchronism of the electrons with this force (the beam wave), i.e.,

$$\omega = (k + |k_0|)v_b, \tag{5.2}$$

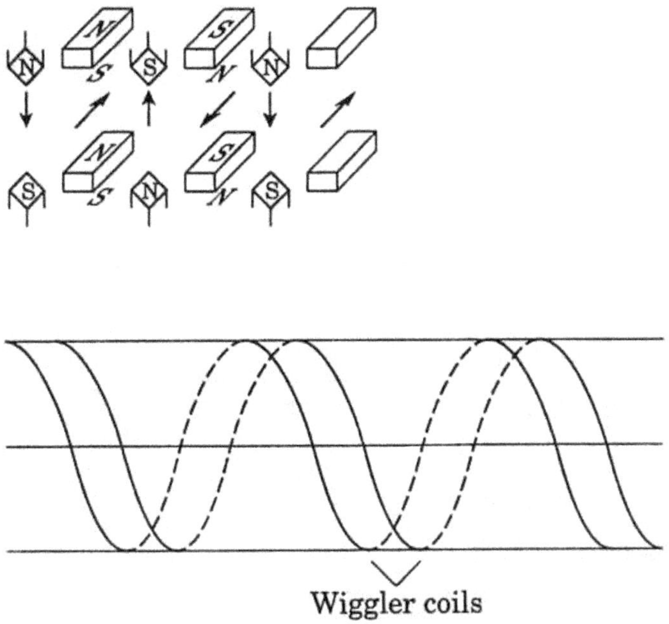

Fig. 5.4. Helical wiggler.

leads to electron bunching and the growth of the radiation wave. Since $\omega = kc$ for the electromagnetic radiation, Eq. (5.2) gives explicitly the frequency of radiation

$$\omega = \frac{|k_0|v_b c}{c - v_b} \simeq 2\gamma_0^2 |k_0| c \,, \tag{5.3}$$

where $\gamma_0 = (1 - v_b^2/c^2)^{-1/2}$. The process is often referred to as stimulated Compton scattering. To the beam electrons the wiggler appears as a backward-propagating electromagnetic wave of frequency $\gamma_0|k_0|v_b$ and wave vector $-\gamma_0|k_0|\hat{z}$. Compton back scattering of this wave produces a wave of frequency $\gamma_0|k_0|v_b$ and wave vector $\gamma_0|k_0|\hat{z}$ which in the laboratory frame has the frequency given by Eq. (5.3). In Fig. 5.5 we have plotted the light line $\omega = kc$, and the beam line $\omega = kv_b$. When an electron loses $\hbar k\ \hat{z}$ momentum it must also lose $\hbar\omega = \hbar k\ v_b$ energy to satisfy $\varepsilon = mc(1 + p^2/m^2c^2)^{1/2}$. For any ω, an electron loses more momentum than a photon can take, hence, in free space without a wiggler there is no emission. In a FEL, the difference in momentum between that given by electrons and absorbed by photons is taken up by the wiggler. One may treat the wiggler as virtual photons of zero energy and momentum $\hbar k_0$.

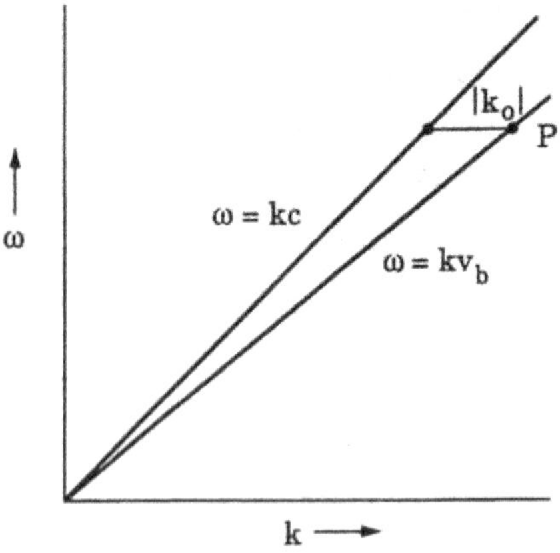

Fig. 5.5. Operating point of a FEL. The electron loses energy $\hbar\omega$, momentum $\hbar k = \hbar\omega/v_b$ producing photon of energy $\hbar k$, momentum $\hbar(k - |k_0|)$. The difference of momentum is shared by the wiggler.

The operating frequency of FEL is obtained by drawing a horizontal line whose length between the light line and the beam mode is k_0. The most attractive feature of the device is that its frequency scales as γ_0^2. It can be tuned over a wide frequency range by changing beam energy.

5.2. Cerenkov FEL

A schematic of a Cerenkov free electron laser is shown in Fig. 5.6. It comprises an electron accelerator (e.g., a Marx generator/Blumlein assembly and a field emission diode giving beam current ≥ 1 Amp and beam voltage ≥ 0.5 MV), a partially dielectric loaded waveguide, a beam collector (or an energy recovery system), a radiation horn or a coupler and the diagnostics. We focus our attention on the physics of the interaction region. For the sake of mathematical simplicity we model the interaction region as an infinite dielectric of effective permittivity ε through which a uniform electron beam of density n_0 propagates with a velocity $v_b\hat{z}$. An electromagnetic signal also exists in this region

$$\mathbf{E} = \mathbf{E}_0 e^{-i(\omega t - \mathbf{k}\cdot\mathbf{x})}, \qquad \mathbf{B} = c\mathbf{k} \times \frac{\mathbf{E}}{\omega}, \tag{5.4}$$

where $\mathbf{E}_0$ and $\mathbf{k}$ lie in the x-z plane. The response of beam electrons to the signal is governed by the relativistic equation of motion

$$\frac{\partial}{\partial t}(\gamma \mathbf{v}) + \mathbf{v} \cdot \nabla(\gamma \mathbf{v}) = -\frac{e}{m}\left(\mathbf{E} + \frac{\mathbf{v}}{c} \times \mathbf{B}\right) \tag{5.5}$$

where $-e$ and m are electronic charge and rest mass. Expanding $\mathbf{v} = \mathbf{v}_b + \mathbf{v}_1$, $\gamma = \gamma_0 + \gamma_0^3 \mathbf{v}_b \cdot \mathbf{v}_1/c^2$ Eq. (5.5) can be linearized to obtain

$$\gamma_0 \mathbf{v}_1 + \frac{\gamma_0^3 v_b^2}{c^2} v_{1z}\hat{z} = \frac{e\left[\mathbf{E}\left(1 - \frac{k_z v_b}{\omega}\right) + \mathbf{k}\frac{v_b}{\omega}E_z\right]}{mi(\omega - k_z v_b)} \tag{5.6}$$

where m is the rest mass and $\gamma_0 = (1 - v_b^2/c^2)^{-1/2}$. Using v_{1x} and v_{1z} in the linearized equation of continuity (with $n = n_0 + n_1$) we obtain

$$\begin{aligned} n_1 &= n_0 \frac{\mathbf{k} \cdot \mathbf{v}_1}{(\omega - k_z v_b)} \\ &= \frac{n_0 e}{mi\gamma_0 \omega(\omega - k_z v_b)}\left[k_x E_x + \frac{k_x^2 v_b E_z}{\omega - k_z v_b} + \frac{\omega k_z E_z}{\gamma_0^2(\omega - k_z v_b)}\right] . \end{aligned} \tag{5.7}$$

The perturbed current density is

$$\begin{aligned} \mathbf{J}_1 &= -n_1 e \mathbf{v}_0 - n_0 e \mathbf{v}_1 \\ &= -\hat{x}\frac{n_0 e^2}{mi\gamma_0\omega}\left(E_x + \frac{k_x v_b}{\omega - k_z v_b}E_z\right) - \hat{z}\frac{n_0 e^2}{mi\gamma_0\omega(\omega - k_z v_b)} \\ &\times \left[k_x v_b E_x + \frac{k_x^2 v_b^2 E_z}{\omega - k_z v_b} + \frac{\omega k_z v_b E_z}{\gamma_0^2(\omega - k_z v_b)} - \frac{\omega}{\gamma_0^2}E_z\right] \end{aligned} \tag{5.8}$$

Using Eq. (5.8) in the wave equation,

$$k^2\mathbf{E} - \mathbf{k}(\mathbf{k} \cdot \mathbf{E}) - \frac{\omega^2}{c^2}\varepsilon \mathbf{E} = \frac{4\pi i \omega}{c^2}\mathbf{J}_1$$

and writing x and z components of the latter we obtain

$$\left(k_z^2 - \frac{\omega^2}{c^2}\varepsilon + \frac{\omega_p^2}{\gamma_0 c^2}\right)E_x = \left(k_x k_z - \frac{\omega_p^2}{\gamma_0 c^2}\frac{k_x v_b}{\omega - k_z v_b}\right)E_z ,$$

$$\begin{aligned} &\left[k_x^2 - \frac{\omega^2}{c^2}\varepsilon + \frac{\omega_p^2}{c^2(\omega - k_z v_b)^2\gamma_0^3}\{k_x^2 v_b^2 \gamma_0^2 + \omega k_z v_b - \omega(\omega - k_z v_b)\}\right] \\ &\cdot E_z = \left(k_x k_z - \frac{\omega_p^2}{c^2}\frac{k_x v_b}{\omega - k_z v_b}\right)E_x . \end{aligned} \tag{5.9}$$

Equations (5.8) and (5.9) give the dispersion relation

$$\left(k_z^2 - \frac{\omega^2}{c^2}\varepsilon + \frac{\omega_{\rm p}^2}{\gamma_0 c^2}\right)\left[k_x^2 - \frac{\omega^2}{c^2}\varepsilon + \frac{\omega_{\rm p}^2(k_x^2 v_{\rm b}^2\gamma_0^2 + 2\omega k_z v_{\rm b} - \omega^2)}{\gamma_0^3 c^2(\omega - k_z v_{\rm b})^2}\right]$$

$$= \left(k_x k_z - \frac{\omega_{\rm p}^2}{\gamma_0 c^2}\frac{k_z v_{\rm b}}{(\omega - k_z v_{\rm b})}\right)^2 , \tag{5.10}$$

where $\omega_{\rm p}^2 = 4\pi n_0 e^2/m$. Equations (5.10) can be rearranged by taking the $\omega_{\rm p}^2$ terms to the right hand side, and retaining only those terms which have a resonance denominator $(\omega - k_z v_{\rm b})^2$:

$$\left(\omega^2 - \frac{k^2 c^2}{\varepsilon}\right)(\omega - k_z v_{\rm b})^2 = \frac{k_x^2\omega_{\rm p}^2}{\omega^2\gamma_0^3\varepsilon}(k_x^2 v_{\rm b}^2\gamma_0^2 + \omega^2) . \tag{5.11}$$

The two factors on the left hand side when equated to zero give electromagnetic and beam modes respectively. Let ω_r be the frequency at which both factors are simultaneously zero. To obtain the solution of Eq. (5.11) when $\omega_{\rm p}$ is small we expand ω as $\omega = \omega_r + \delta$. Then Eq. (5.11) gives

$$\delta^3 = Q\exp(2in\pi)$$

or

$$\delta = Q^{1/3}\left(\cos\frac{2n\pi}{3} + i\sin\frac{2n\pi}{3}\right)$$

where

$$Q = \left[\frac{k_x^2\omega_{\rm p}^2(k_x^2 v_{\rm b}^2\gamma_0^2 + \omega^2)}{\omega^2\gamma_0^2\left(2\omega\varepsilon + \omega^2\dfrac{\partial\varepsilon}{\partial\omega}\right)}\right]^{1/3} \tag{5.12}$$

$n = 0, 1, 2$ and we have dropped subscript r on ω. For $n = 1, \delta$ has a positive imaginary part, giving the growth rate γ of the wave

$$\gamma = \operatorname{Im}\delta = Q^{1/3}\frac{\sqrt{3}}{2} . \tag{5.13}$$

The growth rate γ should not be confused with the relativistic gamma factor. δ has a finite real part also $\delta_r = -\gamma/\sqrt{3}$, i.e., $\omega = k_z v_{\rm b} - \gamma/\sqrt{3}$ or $v_{\rm b} > \omega/k_z$. This is the essential condition for electron bunching and net transfer of energy from electrons to the wave, as explained earlier. The growth rate vanishes when $k_x = 0$, i.e., when the wave propagates parallel to the beam because in this case $E_z = 0$ and bunching of electrons does

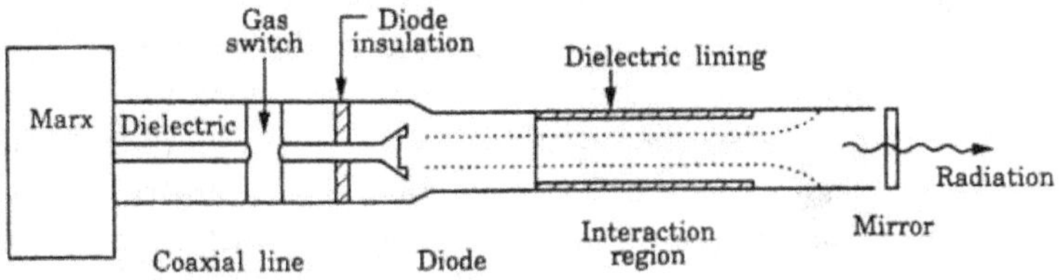

Fig. 5.6. Schematic of a Cerenkov free electron laser.

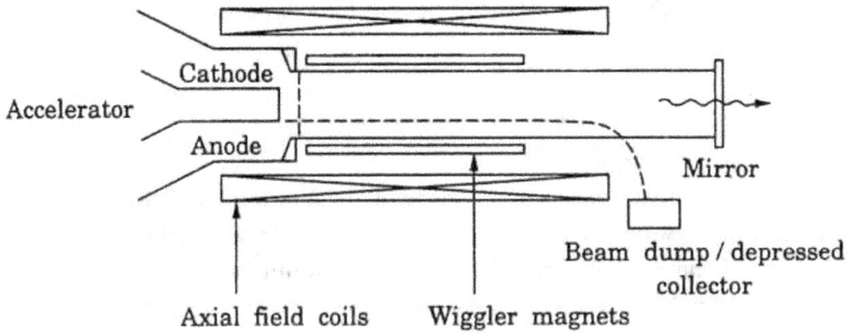

Fig. 5.7. Schematic of a free electron laser.

not take place. k_x (or k_1) in a dielectric loaded waveguide is fixed by the mode number and the geometry of the waveguide.

The nonlinear saturation of Cerenkov instability is a difficult problem to analyze. However, an upper bound on the efficiency can obtained as follows. As the beam loses energy to the wave its velocity decreases. When v_b equals ω/k_z, growth of the instability must stop. The maximum energy available to the radiation is thus $\Delta\varepsilon = mc^2[(1-v_0^2/c^2)^{-1/2}-(1-\omega^2/k_z^2c^2)^{-1/2}]$, and the efficiency of energy conversion

$$\eta \leq \frac{|\delta|}{\omega}\frac{\gamma_0^3}{\gamma_0-1} \, . \tag{5.14}$$

5.3. Free Electron Laser

A geometrical sketch of a free electron laser (FEL) is shown in Fig. 5.7. The foremost component is an accelerator that produces an electron beam with beam voltage $\geq$ 500 kV (could be as high as 1 GV) and beam current $\geq$ 1 A. (up to 100 kA). For beam propagation in a vacuum the beam current is limited to 17 kA due to the space charge effects.[12] To allow for higher beam current one must introduce a plasma in the beam propagation channel to provide charge and partial current neutralization.[13] The beam is usually pulsed with life time $\leq$ 100 nsec. In some experiments one may have pulse lengths $\sim$ 1 msec. The electrons are accelerated in the diode structure which has a hot or a cold cathode together with focusing elements and/or a guide magnetic field. Since the FEL frequency sensitively depends on γ_0, the energy spread of the beam is to be kept small. The beam radius is typically 1 cm and beam thickness $\sim$ 1–2 mm. The interaction region is comprised of a guide magnetic field (produced by a solenoid) to guide the beam and a wiggler magnetic field produced by a suitable arrangement of permanent magnets or by a coil. The typical value of the wiggler strength $\sim$ 1 kG and the wiggler period $\simeq$ 3 cm. Simple wigglers have B_0 and k_0 constant along the length of the interaction region. One may also have a tapered wiggler whose amplitude B_0 and period $(2\pi/k_0)$ change adiabatically along the structure. The latter leads to adiabatic slowing down of the ponderomotive wave, hence, of the trapped electrons leading to improved efficiency of the device. An FEL oscillation would have two reflecting mirrors with broad band high reflectance. Energy recovery from the spent beam is another important consideration. It could be accomplished by using depressed collectors. Here we restrict our discussion to the growth of the FEL radiation in the interaction region, ignoring boundary effects.

Growth rate

Let us consider the propagation of a relativistic electron beam through a wiggler magnetic field

$$\mathbf{B}_0 = (\hat{x} + i\hat{y})B_0 e^{ik_0 z} , \tag{5.15}$$

where $k_0 = -|k_0|$ and the real part of the RHS is implied. The equilibrium electron velocity $\mathbf{v}_0$ is governed by the equation of motion:

$$\frac{d}{dt}(\gamma_0 \mathbf{v}_0) = -\frac{e}{mc}\mathbf{v}_0 \times (\hat{x} + i\hat{y})B_0 e^{ik_0 z} . \tag{5.16}$$

Since γ_0 is a constant for electron motion in a magnetic field, Eq. (5.16) can be solved easily by replacing d/dt by $v_{0z}d/dz = ik_0 v_{0z}$:

$$v_{0x} = \frac{eB_0 e^{ik_0 z}}{mc\gamma_0 k_0}, \qquad v_{0y} = iv_{0x}, \qquad v_{0z} = v_b \ ,$$

$$\gamma_0 = \left[1 - \frac{v_b^2}{c^2} - \frac{e^2 B_0^2}{m^2 c^2 \gamma_0^2 k_0^2 c^2}\right]^{-1/2} . \tag{5.17}$$

Here we have ignored the guide field $B_s \hat{z}$ which is valid as long as $\gamma_0 k_0 v_b > eB_s/mc$.

Perturb this equilibrium by an electromagnetic wave

$$\mathbf{E}_1 = \mathbf{A}_1 e^{-i(\omega_1 t - k_1 z)}, \qquad \mathbf{B}_1 = ck_1 \times \frac{\mathbf{E}_1}{\omega_1} . \tag{5.18}$$

The wave produces an oscillatory electron velocity

$$v_1 = \frac{eE_1}{im\gamma_0\omega_1} \tag{5.19}$$

and exerts a ponderomotive force $\mathbf{F}_{2P} = -(e/2c)(\mathbf{v}_0^* \times \mathbf{B}_1 + \mathbf{v}_1 \times \mathbf{B}_0^*) = e\nabla\phi_p$ on them, with ponderomotive potential

$$\phi_p = -\frac{eB_0(A_{1x} - iA_{1y})}{im\omega k_0 c\gamma_0} e^{-i(\omega t - kz)} \tag{5.20}$$

where $\omega = \omega_1$, $k = k_1 + |k_0|$. The ponderomotive force produces oscillatory electron velocity $v_z \hat{z}$ and density perturbation n having phase variation $e^{-i(\omega t - kz)}$. Solving the relativistic equation of motion and the equation of continuity we obtain

$$v_z = -\frac{ek_z\phi_p}{m\gamma_0^3(\omega - kv_b)}, \qquad n = -\frac{n_0 ek^2\phi_p}{m\gamma_0^3(\omega - kv_b)^2} \tag{5.21}$$

The current density at ω_1, $\mathbf{k}_1$ can now be written as

$$\begin{aligned}\mathbf{J}_1 &= -n_0 e\mathbf{v}_1 - \frac{1}{2}n_2 e\mathbf{v}_0 \\ &= -\frac{n_0 e^2 \mathbf{E}_1}{mi\gamma_0\omega} + \frac{ie^2k^2(E_{1x} - iE_{1y})}{2m\gamma_0^3(\omega - kv_b)^2}\frac{n_0 e^2 B_0^2(\hat{x} + i\hat{y})}{m^2c^2k_0^2\gamma_0^2\omega} .\end{aligned} \tag{5.22}$$

Using $\mathbf{J}_1$ in the wave equation

$$k_1^2 \mathbf{E}_{1\perp} - \frac{\omega_1^2}{c^2}\mathbf{E}_{1\perp} = \frac{4\pi i\omega_1}{c^2}\mathbf{J}_{1\perp} , \tag{5.23}$$

we obtain the nonlinear dispersion relation

$$(\omega^2 - k_1^2c^2)(\omega - kv_b)^2 = \frac{k^2|v_{0\perp}|^2}{4}\frac{\omega_p^2}{\gamma_0^3} \tag{5.24}$$

where $\omega_p^2 = 4\pi n_0 e^2/m$. Around simultaneous zeros of the factors on the left hand side, $\omega_r - k_1 c = 0$, $\omega_r - kv_b = 0$ (giving radiation and beam modes respectively) we expand $\omega = \omega_r + \delta$ to obtain the growth rate

$$\Gamma = \mathrm{Im}\,\delta = \left[\frac{|v_{0\perp}|^2}{c^2}\frac{\omega_p^2 2|k_0|c}{\gamma_0}\right]^{1/3}\frac{\sqrt{3}}{2}\,. \tag{5.25}$$

The growth rate Γ should not be confused with the relativistic gamma factor and $\delta_r = -\Gamma/\sqrt{3}$. The growth rate scales as $B_0^{2/3}$, $k_0^{-1/3}$, Γ_0^{-1}, $n_0^{1/3}$. For fixed k_0, Γ goes as $\omega_1^{-1/2}$, ω_1 is the frequency of radiation. For typical parameters $B_0 = 1$ kG, $2\pi k_0^{-1} = 3$ cm, current density $\simeq 10$ A/cm^2, $\gamma_0 = 4$, i.e., $n_0 = 2 \times 10^9$ cm^{-3}, $\omega_p = 2.2 \times 10^9$ rad sec^{-1}, $eB_0/mk_0c^2 = 0.26$ we obtain $\omega_1 = 320$ GHz, $\Gamma \simeq 2 \times 10^9$ sec^{-1} or spatial growth length $c/\Gamma \simeq 5$ cm.

So far we have ignored the space charge field produced by charge bunching n, which is reasonable at low beam densities. However, at high beam density (i.e., at high beam current), when the frequency of collective oscillations approaches the beat wave frequency, the space charge field becomes important, shifting the FEL operation from the Compton to Raman regime.

Raman regime operation

First, a word about the space charge waves. Consider an uniform beam of cold electrons of density n_0 and velocity $v_b\hat{z}$ subjected to an electrostatic perturbation $\mathbf{E} = -\nabla\phi$; $\phi \sim e^{-i(\omega t - kz)}$. Solving the equations of motion and continuity one may write the density perturbation

$$n_2 = -\frac{n_0 e k^2 \phi}{m\gamma_0^3(\omega - kv_b)^2} \tag{5.26}$$

which on using in the Poisson equation $\nabla^2\phi = 4\pi e n$ yields

$$\varepsilon\phi = 0$$

where

$$\varepsilon = 1 + \chi_b = 1 - \frac{\omega_p^2}{\gamma_0^3(\omega - kv_b)^2}\,. \tag{5.27}$$

$\varepsilon = 0$ gives the space charge modes

$$\omega = kv_b \pm \frac{\omega_p}{\gamma_0^{3/2}} . \tag{5.28}$$

The one with the lower sign has $\omega\partial\varepsilon/\varepsilon\omega < 0$, i.e., it is a negative energy mode. It is the coupling of a negative energy mode $(\omega, \mathbf{k})$ with the wiggler $(0, \mathbf{k}_0)$ that produces the FEL radiation $(\omega_1, \mathbf{k}_1)$ in a Raman regime. As the space charge mode feeds energy to the FEL mode, its energy becomes more negative, leading to the simultaneous growth of the beam space charge mode and the radiation mode.

In the presence of space charge potential ϕ and ponderomotive potential ϕ_p, the density perturbation can be written from Eq. (5.26) with ϕ replaced by $\phi + \phi_p$. The Poisson equation then yields

$$\varepsilon\phi = -\chi_b\phi_p . \tag{5.29}$$

The current density at $(\omega_1, \mathbf{k}_1)$ now takes the form

$$\mathbf{J}_1 = -n_0 e \mathbf{v}_1 - \frac{1}{2} n_2 e \mathbf{v}_0 = -\frac{n_0 e^2 \mathbf{E}}{mi\gamma_0\omega} + \frac{k^2}{8\pi}\phi\mathbf{v}_0 \tag{5.30}$$

and the wave equation (5.23) reads as

$$\left(\omega^2 - \frac{\omega_p^2}{\gamma_0} - k_1^2c^2\right)\mathbf{E}_{1\perp} = -\frac{i\omega}{2}k^2\phi\mathbf{v}_0 . \tag{5.31}$$

Equations (5.29) and (5.31) yield the nonlinear dispersion relation

$$\left(\omega^2 - \frac{\omega_p^2}{\gamma_0} - k_1^2c^2\right)\varepsilon = -\frac{k^2|v_{0\perp}|^2}{4}\chi_b . \tag{5.32}$$

The simultaneous zero of the two factors on the LHS gives the frequency of operation of the FEL

$$\omega^2 = \frac{\omega_p^2}{\gamma_0} + k_1^2c^2,$$

$$\omega = kv_b - \frac{\omega_p}{\gamma_0^{3/2}} \tag{5.33}$$

For $\omega \gg \omega_p/\gamma_0^{1/2}$, $\omega \simeq 2\gamma_0^2(k_0v_b - \omega_p/\gamma_0^{3/2})$. Around this frequency we expand $\omega = \omega + i\Gamma$. Equation (5.32) yields

$$\Gamma = \frac{k|v_{0\perp}|}{4}\left(\frac{\omega_{\rm p}}{\gamma_0^{3/2}\omega}\right)^{1/2} . \tag{5.34}$$

The neglect of collective effects implies $\chi_{\rm b} \ll 1$, i.e., $\omega_{\rm p}/\gamma_0^{3/2} < \Gamma$. This defines the boundary between Raman and Compton regimes of operation.

Nonlinear state

As the amplitude of the space charge wave (or the beat wave) grows it tends to trap the electrons. Trapped electrons, however, cannot continuously be decelerated by the beat wave. To understand the dynamics of trapped electrons in the ponderomotive wave we write the single particle equation of motion neglecting space charge effects,

$$\frac{d\gamma_e}{dz} = -\frac{eE_{\rm p}}{mc^2}\cos(\omega_1 t - kz) \tag{5.35}$$

where

$$\gamma_e = \left(1 + \frac{p_\perp^2 + p_z^2}{m^2c^2}\right)^{1/2} = \left(1 - \frac{v_{0\perp}^2}{c^2}\frac{v_z^2}{c^2}\right)^{-1/2}, \qquad E_{\rm p} = k|\phi_{\rm p}|$$

$\mathbf{p}$ is the electron momentum and we have written $dP_z/dt = v_z dP_z/dz \equiv mc^2 d\gamma_e/dz$. It is useful to define the gamma factor of an electron moving with the phase velocity of the wave $\gamma_r = \left[1 - (v_{0\perp}^2/c^2) - (\omega_1^2/k^2c^2)\right]^{-1/2}$. When γ_e falls to γ_r the beam can no longer transfer energy to the wave. Let $\Delta\gamma_e = \gamma_e - \gamma_r$, $\psi = kz - \omega_1 t$ to write

$$\frac{d}{dz}\Delta\gamma_e = -\frac{eE_{\rm p}}{mc^2}\cos\psi$$

$$\frac{d\psi}{dz} = k - \frac{\omega}{v_z} = \frac{\omega}{c}\left[\left(1 - \frac{1}{\gamma_r^2}\right)^{-1/2} - \left(1 - \frac{1}{\gamma_e^2}\right)^{-1/2}\right]$$

$$\simeq \frac{\omega\Delta\gamma_e}{2c(\gamma_r - 1)^{3/2}} . \tag{5.36}$$

Dimensionalizing z by the length of the interaction region L, i.e., $\xi = z/L$, and defining

$$P = \frac{L\omega\Delta\gamma_e}{2c(\gamma_r^2 - 1)^{3/2}}, \qquad A = \frac{L^2\omega eE_{\rm p}}{2mc^3(\gamma_r^2 - 1)^{3/2}}$$

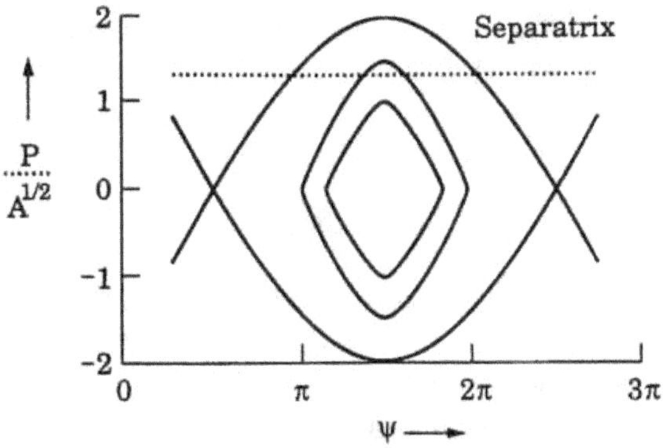

Fig. 5.8. Phase space trajectories of trapped particles.

Eqs. (5.36) can be cast as

$$\frac{dP}{d\xi} = -A\cos\psi, \tag{5.37}$$

$$\frac{d\psi}{d\xi} = P\ . \tag{5.38}$$

This set is to be solved in conjunction with the equation governing A, i.e., the wave equation for the radiation wave. Its solution, however, can be obtained only numerically. Here we make some general observations. An electron can lose energy to the wave as long as ψ is between $-\pi/2$ and $\pi/2$. As the electron moves down the interaction region ψ could increase (for $P > 0$). One would like to choose a length of interaction region such that the change in ψ at the exit point ($\xi = 1$) is π, i.e., $P_{\text{exit}} - P_{\text{entry}} \simeq \pi$. Hence the efficiency of energy conversion in a FEL is

$$\eta - \frac{\gamma_e(\xi = 0) - \gamma_e(\xi = 1)}{\gamma_e(\xi = 0) - 1} \simeq \frac{\pi}{k_0 L}\ .$$

Gain estimate

It is worthwhile solving Eqs. (5.37)–(5.38) in the limit when A is constant, i.e., the single pass amplification of radiation is small. Dividing Eq. (5.37) by (5.38) and integrating the resulting equation we obtain

$$P^2 = -2A\sin\psi + P_{\text{in}}^2 + 2A\sin\psi_{\text{in}}\ , \tag{5.39}$$

where P_{in} and ψ_{in} are the values of P and ψ at the entry point $z = 0$. Equation (5.39) gives phase (P, ψ) space trajectories of the electron (Fig. 5.8) for different values of $P_{\text{in}}, \psi_{\text{in}}$. For $P_{\text{in}}^2 + 2A \sin \psi_{\text{in}} < 2A$, all values of ψ are not accessible (since P^2 has to be > 0), i.e., the trajectories of particles are localized, representing trapped particles. The separatrix is given by

$$P^2 = 2A(1 - \sin \psi) . \tag{5.40}$$

At $z = 0$, the electrons enter uniformly at all times, i.e., in the P, ψ plane they lie uniformly over the horizontal line $P = P_{\text{in}}$. The ones inside the separatrix are trapped and those outside are the passing particles. At $z > 0$, some lose energy, some gain energy depending on their ψ_{in}. To have a quantitative estimate of net energy exchange we solve Eqs. (5.37) and (5.38) by expanding P and ψ to different orders in A

$$P = P_0 + P_1 + P_2, \qquad \psi = \psi_0 + \psi_1 + \psi_2 . \tag{5.41}$$

To the zeroeth order:

$$P_0 = P_{\text{in}}, \qquad \psi_0 = \psi_{\text{in}} + P_{\text{in}}\xi .$$

To the first order, from Eqs. (5.39) and (5.38) respectively,

$$\begin{aligned} P_1 &= -\frac{A}{P_{\text{in}}} [\sin(\psi_{\text{in}} + P_{\text{in}}\xi) - \sin \psi_{\text{in}}] \\ \psi_1 &= \frac{A}{P_{\text{in}}^2} [\cos(\psi_{\text{in}} + P_{\text{in}}\xi) - \cos \psi_{\text{in}}] + \xi \frac{A}{P_{\text{in}}} \sin \psi_{\text{in}} . \end{aligned} \tag{5.42}$$

Using $\psi = \psi_0 + \psi_1$ in Eq. (5.39) we get

$$P_2 = \frac{A}{P_{\text{in}}} [\sin \psi_{\text{in}} - \sin \psi_0 - \psi_1 \cos \psi_0] - \frac{P_1^2}{2P_{\text{in}}} .$$

The energy lost by an electron $\Delta P \equiv P_{\text{in}} - P(\xi = 1)$ in passing through the interaction region is $\Delta P = -(P_1 + P_2)_{\xi=1}$. Average ΔP over the initial phases yields

$$\langle \Delta P \rangle = \frac{A^2}{P_{\text{in}}^3} \left[1 - \cos P_0 - \frac{P_0}{2} \sin P_0 \right] = -\frac{A^2}{8} \frac{d}{dx} \left(\frac{\sin^2 x}{x^2} \right) , \tag{5.43}$$

where $x = P_0/2$. Figure 5.9 shows the variation of gain function $G \equiv -(d/dx)\sin^2 x/x^2$ as a function of P_0 or $\gamma_0 - \gamma_r$. For $\gamma_0 > \gamma_r$ there is a net transfer of energy from the particles to the wave.

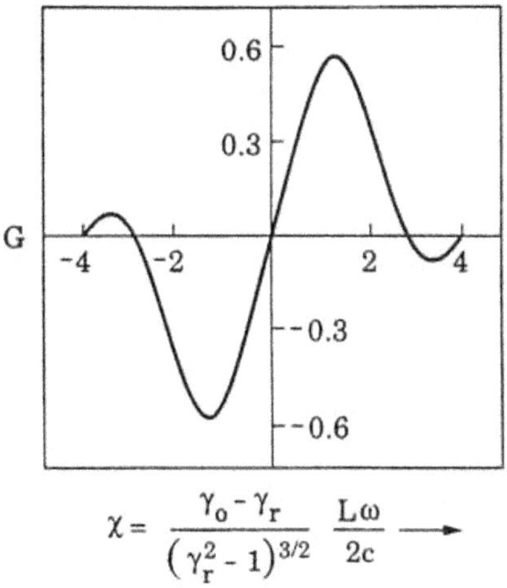

Fig. 5.9. Gain function as a function of initial electron energy [cf. Ref. 1].

At high beam current the amplitude of the FEL wave grows significantly during the passage of an electron through the wiggler, hence one must supplement the set (5.37) and (5.38) with an equation for A. Roberson and Sprangle[2] have discussed this problem in considerable detail. However, we conclude this section with an estimate of efficiency. In the Compton regime of FEL operation electrons give energy to the wave as long as $v_b > \omega/k$. During the emission process, electron velocity thus falls from $v_b = (\omega + \delta r)/k$ to $v_b = \omega/k$, giving efficiency

$$\eta = \frac{\gamma_e\left(\frac{\omega+\delta r}{k}\right) - \gamma_e\left(\frac{\omega}{k}\right)}{\gamma_e\left(\frac{\omega+\delta r}{k}\right) - 1} = \frac{\delta r}{\omega}\frac{\gamma_0^3}{\gamma_0 - 1}\,. \tag{5.44}$$

In the Raman case the frequency $\omega = kv_b - \omega_p/\gamma_0^{3/2}$ of the space change mode is detuned as the beam loses energy. When $k\Delta v_b$ is of the order of the growth rate the instability should saturate. Hence,

$$\eta \sim \frac{\gamma_0^3}{\gamma_0 - 1}\cdot\frac{\gamma}{\omega_1}\,.$$

Tapered wiggler FEL

The efficiency of a free electron laser can be enhanced by adiabatically slowing down the ponderomotive wave so that $v_b > \omega/k(z)$ can be maintained over a long distance. This is accomplished by tapering the wiggler, viz, by making k_0 an increasing function of z. Let us examine the motion of an electron in a ponderomotive wave $E_p = E_p \cos\psi$ where $\psi = \int k(z)dz - \omega_1 t$, $k = k_1 + k_0(z)$. For the sake of simplicity let us assume $v_{0\perp}^2 \ll c^2$. We define a resonant gamma

$$\gamma_r = \left(1 - \frac{\omega_1^2}{k^2(z)c^2}\right)^{-1/2} \tag{5.45}$$

and $\Delta\gamma_e \equiv \gamma_e - \gamma_r$. Then Eq. (5.35) can be written as

$$\frac{d\Delta\gamma_e}{dz} = -\frac{eE_p}{mc^2}\cos\psi - \frac{d\gamma_r}{dz}$$

$$\frac{d\psi}{dz} = \frac{\omega\Delta\gamma_e}{2c(\gamma_r - 1)^{3/2}}$$

Defining ξ, P and A the same way as before and assuming

$$\frac{1}{\gamma_r}\frac{d\gamma_r}{dz} \ll \frac{1}{\Delta\gamma_e}\frac{d}{dz}\Delta\gamma_e$$

the above equations can be written as

$$\frac{dP}{d\xi} = -A\cos\psi + \alpha\,,$$

$$\frac{d\psi}{d\xi} = P\,, \tag{5.46}$$

where $\alpha = -[2c(\gamma_r^2 - 1)^{3/2}/\omega]d\gamma_r/dz$. In general α is a slowly varying function of ξ. We assume α to be a constant. Writing $dP/d\xi = (dP/d\psi)(d\psi/d\xi)$ we get

$$\frac{d}{d\psi}\left(\frac{P^2}{2}\right) = -A\cos\psi + \alpha \tag{5.47}$$

which on integration gives

$$\frac{P^2}{2} + A\sin\psi - \alpha\psi = H \tag{5.48}$$

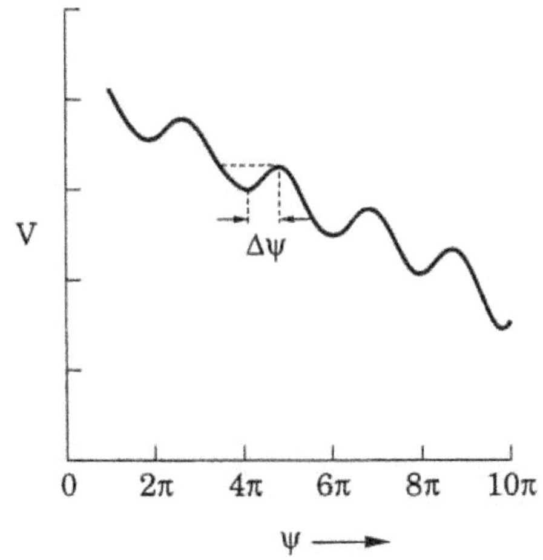

Fig. 5.10. Potential energy as a function of phase angle ψ for $\alpha < A$.

where $H = P_{\text{in}}^2/2 + A\sin\psi_{\text{in}} - \alpha\psi_{\text{in}}$ is a constant of integration. H can be identified as a Hamiltonian, $P^2/2$ as the kinetic energy and

$$V(\psi) = A\sin\psi - \alpha\psi \tag{5.49}$$

as the potential energy.We have plotted V as a function of ψ for $A > \alpha_{\text{in}}$ (Fig. 5.10) It has the form of a sloping corrugated roof, with maxima-minima occurring ($dV/d\psi = 0$) at

$$\cos\psi = \alpha/A\ . \tag{5.50}$$

Equation (5.50) is satisfied when $\alpha < A$. For fast tapering i.e. $\alpha > A$ one would not have maxima-minima. For a minimum $d^2V/d\psi^2 > 0$ i.e., $A\sin\psi_{\text{min}} < 0$. Since $\cos\psi_{\text{min}}$ is positive (cf. Eq. (5.50)) ψ_{min} lies in the fourth quadrant. (cf. Fig. 5.11). The values of ψ ($= \psi_{\text{max}}$) corresponding to maxima in V lie in the first quadrant. The distance in ψ separation between a minimum and the next maximum is $\Delta\psi = 2\cos^{-1}\theta/A$, while the width of a potential energy well is $2\Delta\psi$. For an electron having $H < V_{\text{max}}$, where $V_{\text{max}} \equiv A\sin\psi_{\text{max}} - \alpha\psi_{\text{max}} = A\left(1 - \alpha^2/A^2\right)^{1/2} - \cos^{-1}\alpha/A$, all values of ψ are not accessible because P^2 is a positive definite. The electron bounces back and forth inside a potential energy well. Such electrons are trapped electrons. We have plotted phase space trajectories of electrons in Fig. 5.12, with the separatrix given by

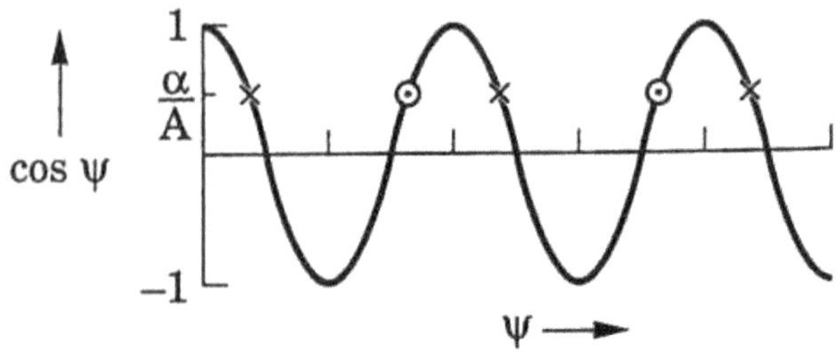

Fig. 5.11. Locations of $\psi_{\min}(0)$ and $\psi_{\max}$; $\Delta\psi$ is the half width of potential energy well.

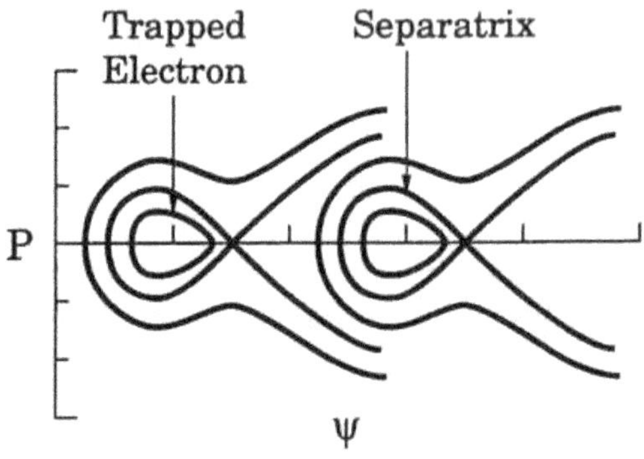

Fig. 5.12. Phase space trajectories of electrons.

$$\frac{P^2}{2} + A\sin\psi - \alpha\psi = (A^2 - \alpha^2)^{1/2} - \alpha\cos^{-1}\frac{\alpha}{A}\,. \tag{5.51}$$

As the ponderomotive wave slows down with z the trapped electrons also slow down losing energy to FEL radiation. Let the values of γ_r and P at $\xi = 0$ and $\xi = 1$ be denoted as $\gamma_r(0)$, $P(0)$ and $\gamma_r(1)$, $P(1)$, then the net energy radiated by trapped electrons is

$$\Delta\varepsilon = mc^2\left[\gamma_r(0) - \gamma_r(1) + \frac{L\omega\langle P(0) - P(1)\rangle}{2c(\gamma_r^2(0) - 1)^{3/2}}\right]$$

where $\langle\ \rangle$ denotes the average over the initial phases for trapped electrons. $\langle P(0) - P(1)\rangle$ can be obtained by solving Eqs. (5.46) numerically.

The efficiency of radiation of trapped electrons is given by

$$\eta_{\rm tr} = \frac{\Delta\varepsilon}{(\gamma_e(0) - 1)mc^2} \tag{5.52}$$

To this if we multiply the fraction of electrons trapped, we would get the overall efficiency of FEL radiation. In case, where the electrons enter the interaction region ($z = 0$) at a uniform rate, i.e., they are distributed uniformly over ψ, then the fraction of those trapped in $2\Delta\psi/2\pi$ and the overall efficiency of FEL radiation is given by

$$\eta = \frac{\Delta\psi}{\pi}\frac{\Delta\varepsilon}{(\gamma_e(0)-1)mc^2}$$

$$\gtrsim \frac{\Delta\psi}{2\pi}\frac{\gamma_r(0)-\gamma_r(1)}{(\gamma_e(0)-1)} \,. \tag{5.53}$$

To the above expression one must add the contribution due to untrapped electrons; however, that is relatively weaker in a tapered free electron laser.

References

1. T. C. Marshall, *Free Electron Lasers* (McMillan, New York, 1985), p. 78.
2. C. W. Roberson and P. Sprangle, *Phys. Fluids* **B1**, 3 (1989); also see reference cited therein.
3. L. S. Bogdankevich, M. V. Kuzelev, and A. A. Rukhadze, *Sov. Phys. Ups.* **24**, 1 (1981).
4. M. V. Kuzelev, A. A. Rukhadze, P. S. Strelkov, and A. G. Shkvarunets, *Sov. J. Plasma Phys.* **13**, 793 (1987).
5. J. E. Walsh, "Cerenkov and Cerenkov-Raman radiation sources, in free electron generators of coherent radiation" *Phys. of Quant. Electron* Vol. 7, eds. S. F. Jacobs, H. S. Pilloff, M. Sargent III, M. O. Scully, and R. Spitzer (Addison-Wesley, 1980), p. 255.
6. H. Motz, *J. Appl. Phys.* **22**, 527 (1951), *ibid. Contemp. Phys.* **20**, 547 (1979).
7. L. R. Elias, W. M. Fairbanks, J. M. J. Madey, H. A. Schwettman, and T. I. Smith, *Phys. Rev. Lett.* **36**, 717 (1976).
8. K. R. Chen, J. M. Dawson, A. T. Lin, and T. Katsouleas, *Phys. Fluids* **B3**, 1270 (1991); also see reference, cited therein.
9. T. M. Antensen and B. Levush, *Phys. Fluids* **B1**, 1097 (1989); *ibid. Phys. Rev. Lett.* **62**, 1488 (1989).
10. V. K. Tripathi and C. S. Liu, *IEEE Trans. Plasma Sci.* **18**, 466 (1990).
11. H. P. Freund and T. M. Antonsen Jr., *Principles of Free Electron Lasers* (Chapman & Hall, 1992).
12. R. B. Millar, *An Introduction to the Physics of Charged Particle Beams* (Plenum, 1982).
13. A. F. Alexandrov, L. S. Bogdankevich, and A. A. Rukhadze, *Principles of Plasma Electrodynamics*, Springer Series in Electrophysics, ed. G. Ecker (Springer-Verlag, New York, 1984), Vol. 9.

CHAPTER 6

SELF-FOCUSING AND FILAMENTATION

The propagation of electromagnetic waves of finite transverse extent, or nonuniform intensity distribution, through plasmas is a problem of practical importance. At low power densities we have seen that diffraction causes divergence of the wave. The picture however, is changed drastically at higher power densities.

The electrons, in the presence of a nonuniform electromagnetic wave, experience a ponderomotive force $\mathbf{F}_p \sim -\nabla|E|^2$, directed opposite to the intensity gradient. As the electrons are displaced the ions follow them due to strong space charge forces, resulting in plasma density depression in the illuminated region (cf. Fig. 6.1), on a time scale $\tau_d \sim r_0/c_s$, where r_0 is the scale length of intensity variation and c_s is the sound speed. In the presence of collisions the electrons are heated on a time scale $\tau_H \sim (\delta\nu)^{-1}$, where δ is the mean fraction of energy lost in a collision and ν is the collision frequency. The nonuniform heating rate due to the intensity variation leads to a thermal pressure gradient pushing out the plasma, causing density depression. The depressed density acts as a duct to guide and self-focus the electromagnetic wave, which further intensifies the wave maximum.[1-6]

Consider such a duct due to an electromagnetic beam with Gaussian intensity distribution (cf. Fig. 6.1). The phase velocity of the wave is minimum on the axis, due to reduced density, (i.e., enhanced index of refraction) and constantly increases always from it. As the wave front advances through such a duct it acquires a curvature and tends to focus. Focusing leads to enhancement of intensity, causing even deeper density depression and so

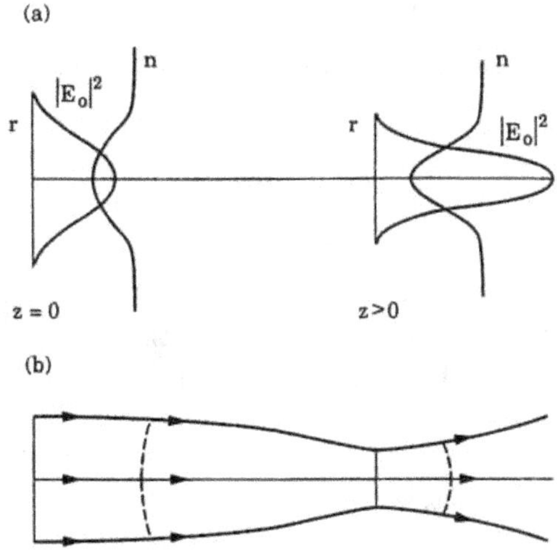

Fig. 6.1. (a) Intensity profiles and density depressions at different values of z. (b) Wave front at different values of z.

on. The focusing continues in a collapsing manner until the duct is nearly depleted of the plasma and diffraction divergence supersedes the nonlinear self-focusing. The beam thus acquires a minimum radius at some point, beyond which it diverges. As beam diameter increases and axial intensity decreases the saturation effect of nonlinearity becomes less important, nonlinear refraction becomes stronger and self-focusing starts. Thus a periodic focusing is realized. It is possible to choose a radiation intensity profile for which self-focusing and diffraction effects are evenly balanced and the beam propagates as a spatial soliton.

Even in an initially uniform electromagnetic wave, a small fluctuation in its intensity distribution along the wave front may be amplified by the same process. This is called the filamentation instability.[7-8]

At very high power densities, when electron oscillatory velocity approaches the velocity of light, the relativistic mass increases with wave intensity, causing a local enhancement in the index of refraction. For a Gaussian beam this produces a refractive index profile suitable for self-focusing. Since the onset of this instability is instantaneous, it is important in short pulse experiments where the other two mechanisms are not operative.

6.1. Nonlinear Permittivity

The effective permittivity of a plasma for the propagation of an electromagnetic wave,

$$\mathbf{E} = \mathbf{A}(\mathbf{x})e^{-i(\omega t - kz)} ,$$

$$\mathbf{B} = c\nabla \times \mathbf{E}/i\omega$$

is given by $\varepsilon = 1 - 4\pi ne^2/m'\omega^2$ where n is the electron density and m' the local electron mass. In the non-relativistic case $m' = m$, and the nonlinearity arises through the modification of n due to ponderomotive force and nonuniform ohmic heating. The time scales for ambipolar motion and ohmic heating are: $\tau_d \sim r_0/c_s$, $\tau_H \sim (\delta\nu)^{-1}$; usually $\tau_d < \tau_H$. The relativistic mass nonlinearity arises instantaneously. For short pulses of duration $\tau \ll \tau_d$, τ_H only the latter nonlinearity is important. It is characterized by a nonlinearity parameter v_0/c where v_0 is the electron oscillatory velocity. For long pulses $\tau_d < z < z_H$ the ponderomotive nonlinearity characterized by $v_0/v_{\rm th}$ is predominant, whereas, for very long pulses $\tau_d, \tau_H < \tau$ the ohmic heating nonlinearity characterized by $v_0/v_{\rm th}\delta^{1/2}$ prevails upon the others. Below, we obtain the quasi steady state expression for permittivity in the three cases.

Relativistic mass nonlinearity

The solution of the relativistic equation of motion for a circularly polarized wave $\mathbf{E} = (\hat{x} - i\hat{y})E_0(x)e^{-i(\omega t - kz)}$, having $\nabla E_0 \ll kE_0$, can be written as

$$\mathbf{v} = \frac{e\mathbf{E}}{mi\omega\gamma_0},$$

$$\gamma_0 = \left(1 + \frac{e^2|E_0|^2}{m^2\omega^2c^2}\right)^{1/2} . \tag{6.1}$$

where γ_0 is the relativistic gamma factor and m is the rest mass of the electron. The permittivity can be written as

$$\varepsilon = \varepsilon_0 + \Phi \tag{6.2}$$

where

$$\varepsilon_0 = 1 - \frac{\omega_p^2}{\omega^2} ,$$

$$\Phi = \frac{\omega_p^2}{\omega^2} \left(1 - \frac{1}{\left(1 + \frac{e^2 |E_0|^2}{m^2 \omega^2 c^2}\right)^{1/2}} \right) . \tag{6.3}$$

One may notice that the characteristic nonlinearity parameter is $eE_0/m\omega c$. In cases where E_0 is uniform with a small perturbation: $E_0 = A_0 + A_1(\mathbf{x})$, ε can be written as Eq. (6.2) with

$$\varepsilon_0 = 1 - \frac{\omega_p^2}{\omega^2 \gamma_0}$$

$$\Phi = \frac{\omega_p^2}{2\omega^2 \gamma_0^3} \frac{e^2 A_0 (A_1 + A_1^*)}{m^2 \omega^2 c^2}$$

$$\gamma_0 = [1 + e^2 A_0^2 / m^2 \omega^2 c^2]^{1/2} . \tag{6.4}$$

Ponderomotive force nonlinearity

This nonlinearity is important in long pulse experiments where power densities are usually low, hence v_0/c is small and relativistic effects can be ignored. Then the equation of motion for electrons can be written as

$$m \frac{\partial \mathbf{v}}{\partial t} = -e\mathbf{E} + \mathbf{F}_p - \frac{1}{n} \nabla (n T_e) , \tag{6.5}$$

where

$$\mathbf{F}_p \equiv -m \mathbf{v} \cdot \nabla \mathbf{v} - \frac{e}{c} \mathbf{v} \times \mathbf{B}$$

is the ponderomotive force. The high frequency response is

$$\mathbf{v} = \frac{e\mathbf{E}}{mi\omega} . \tag{6.6}$$

Using $\mathbf{v}$ and $\mathbf{B}$, the static ponderomotive force can be written as

$$\mathbf{F}_p = -\frac{m}{2} \mathbf{v}^* \cdot \nabla \mathbf{v} - \frac{e}{2c} \mathbf{v} \times \mathbf{B}^* = -\frac{e^2}{2m\omega^2} \nabla (E_0 E_0^*) , \tag{6.7}$$

where we have used the identity $\mathrm{Re}\, A \cdot \mathrm{Re}\, B \equiv \frac{1}{2} \mathrm{Re}[AB + AB^*]$. In the quasi-steady state the net static forces on electron and ion fluids must vanish,

$$-e\mathbf{E}_s + \mathbf{F}_p - \frac{1}{n}\nabla(nT_e) = 0$$

$$ze\mathbf{E}_s - \frac{1}{n_i}\nabla(n_iT_i) = 0 \tag{6.8}$$

where $\mathbf{E}_s$ is the space charge field. Under the condition of quasineutrality $n \simeq zn_i$, Eqs. (6.8) combine to give

$$\left(T_e + \frac{T_i}{z}\right)\nabla \ln n = -\frac{e^2}{2m\omega^2}\nabla(E_0E_0^*) \; . \tag{6.9}$$

We solve this equation in two distinct cases of interest:

Case (i) Uniform beam with an imposed ripple:

$$\mathbf{E}_0 = \mathbf{A}_0 + \mathbf{A}_1(\mathbf{x}) \; ,$$

$$n \simeq n_0^0\left[1 - \frac{e^2\mathbf{A}_0\cdot(\mathbf{A}_1 + \mathbf{A}_1^*)}{2m\omega^2\left(T_e + \frac{T_i}{z}\right)}\right] \; , \tag{6.10}$$

where $A_1 \ll A_0$, A_0 is taken to be real.

Case (ii) A beam of finite extent (e.g., a Gaussian beam)

$$n = n_0^0 e^{-v_{osc}^2/v_{th}'^2} \tag{6.11}$$

where $v_{osc} = e|E_0|/m\omega$, $v_{th}'^2 = 2(T_e + T_i/z)/m$. Density distribution (6.11) is essentially the Boltzmann distribution $n = n_0^0 e^{e\phi_p/T}$, where $\phi_p = \frac{1}{2}mv_{osc}^2$ is the ponderomotive potential ($\mathbf{F}_p = e\nabla\phi_p$). The high-frequency permittivity of the plasma $\varepsilon = 1 - 4\pi ne^2/m\omega^2$ may now be written as Eq. (6.2) with

$$\Phi = \begin{cases} \dfrac{\omega_p^2}{\omega^2}\dfrac{e^2\mathbf{A}_0\cdot(\mathbf{A}_1 + \mathbf{A}_1^*)}{2m\omega^2\,(T_e + T_i/z)} & \text{case (i)} \\[2ex] \dfrac{\omega_p^2}{\omega^2}\left(1 - e^{-v_{osc}^2/v_{th}'^2}\right) & \text{case (ii).} \end{cases} \tag{6.12}$$

In the case of a finite extent beam ε shows a saturation behavior for $v_{osc} > v_{th}'$.

Ohmic nonlinearity

When the time duration of the electromagnetic beam is longer than a temperature relaxation time, ohmic heating of electrons becomes important.

Solving the equation of motion for the oscillatory electron velocity due to the electromagnetic wave we get

$$\mathbf{v} = \frac{e\mathbf{E}}{mi\omega}(1 - i\nu/\omega) \ . \tag{6.13}$$

The component of $\mathbf{v}$ in phase with $\mathbf{E}$ causes electron heating at the rate $-(e/2)\mathbf{E}^* \cdot \mathbf{v} = e^2 E_0 E_0^* \nu / 2m\omega^2$. In the steady state this rate is balanced by the power loss via thermal conduction and collisions with ions and neutrals:

$$-\nabla \cdot \left(\frac{\chi}{n} \nabla T_e\right) + \frac{3}{2}\delta\nu(T_e - T_i) = \frac{e^2 \nu E_0 E_0^*}{2m\omega^2} \ , \tag{6.14}$$

where $\chi/n = v_{th}^2/\nu$, $\delta = 2(m/m_i)$ for electron-ion energy exchange collision. We may define characteristic times for thermal conduction and collisional energy transfer, τ_{con} and τ_{coll}

$$\tau_{con} \sim \frac{\nu r_0^2}{v_{th}^2} \, , \qquad \tau_{coll} \sim (\delta\nu)^{-1}$$

where r_0 is the characteristic scale length of variation of EE^*. For $\tau_{coll} \ll \tau_{con}$ i.e., $\nu^2 r_0^2 \delta / v_{th}^2 > 1$ one may ignore the first term in Eq. (6.13). Then[9]

$$T_e = T_i + \frac{m}{3}\left(\frac{e|E_0|}{m\omega}\right)^2 \frac{1}{\delta} \ . \tag{6.15}$$

One must compare the thermal pressure gradient force $-\nabla T_e \equiv -(1/3\delta) \times (e^2/m\omega^2)\nabla|E_0|^2$ with the ponderomotive force $\mathbf{F}_p$. The former is δ^{-1} times the latter. Since δ is small, the thermal pressure gradient force prevails upon the other one in the steady state. Assuming quasi-neutrality and demanding the uniformity of plasma pressure $n(T_e + T_i) =$ constant, one obtains the modified density leading to nonlinear permittivity:

$$\Phi = \frac{\omega_p^2}{\omega^2} \frac{e^2 \mathbf{A}_0 \cdot (\mathbf{A}_1 + \mathbf{A}_1^*)}{3m\omega^2 \delta (T_i + T_{eo})} \qquad \text{case (i)}$$

$$= \frac{\omega_p^2}{\omega^2} \frac{e^2|A|^2/6\delta\omega^2 v_{th}'^2 m^2}{1 + e^2|E_0|^2/6\delta m^2 v_{th}'^2} \qquad \text{case (ii)} \ . \tag{6.16}$$

Ohmic nonlinearity with thermal conduction

When $\tau_{con} < \tau_{coll}$ thermal conduction is important. In this case, however, Eq. (6.14) can be solved only when $\mathbf{x}$ dependence of $\mathbf{A}$ is known. For case (i) if one takes $A \sim e^{i\mathbf{q}\cdot\mathbf{x}}$ with $q_\parallel \ll q_\perp$ then Eq. (6.14) gives

$$T_e - T_i = \frac{e^2[\mathbf{A}_0 \cdot (\mathbf{A}_1 + \mathbf{A}_1^*) + A_0^2]}{3m\omega^2\delta'}$$

$$n = n_0\left(1 - \frac{e^2\mathbf{A}_0 \cdot (\mathbf{A}_1 + \mathbf{A}_1^*)}{3m\omega^2\delta'(T_i + T_{eo})}\right)$$

$$\Phi = \frac{\omega_p^2}{\omega^2}\frac{e^2\mathbf{A}_0 \cdot (\mathbf{A}_1 + \mathbf{A}_1^*)}{3m\omega^2\delta'(T_i + T_{eo})}, \qquad \delta' = \delta + \frac{2q^2v_{th}^2}{3\nu^2} . \tag{6.17}$$

For a beam of finite extent Eq. (6.14) can be solved analytically only in the weak nonlinearity approximation: $T_e = T_i + \Delta T$, $\Delta T \ll T_i$. Then Eq. (6.14) gives

$$\nabla^2(\Delta T_e) - \frac{3}{2}\frac{\delta\nu^2}{v_{th}^2}(\Delta T_e) = -\frac{e^2\nu^2}{2m\omega^2v_{th}^2}|E_0|^2 . \tag{6.18}$$

Employing pressure balance, $\Delta n = -n_0\Delta T_e/(T_{eo} + T_i) \approx -n_0\nabla T_e/2T_i$, Eq. (6.18) can be recast as

$$\nabla^2(\Delta n) - \frac{3}{2}\frac{\delta\nu^2}{v_{th}^2}(\Delta n) = +\frac{e^2\nu^2 n_0}{2m\omega^2v_{th}^2 2T_i}|E_0|^2 . \tag{6.19}$$

If one employs a Gaussian Ansatz for $|E_0|^2$, as we shall do later,

$$|E_0|^2 = A_0^2 e^{-r^2/r_0^2}$$

Eq. (6.19) yields

$$\Delta n = -\frac{\alpha}{k_0}\left[K_0(k_0r)\int_0^r k_0r'I_0(k_0r')e^{-r'^2/r_0^2}dr'\right.$$

$$\left.+I_0(k_0r)\int_r^\infty k_0r'K_0(k_0r')e^{-r'^2/r_0^2}dr'\right] , \tag{6.20}$$

where $\alpha = e^2\nu^2n_0^0A_0^2/2m^2\omega^2v_{th}^4$, $k_0^2 = 3\nu^2\delta/2v_{th}^2$ and K_0 and I_0 are the modified Bessel functions. From (6.20),

$$\Delta n\Big|_{r=0} = -\frac{\alpha}{k_0^2}I_0(0)\int_0^\infty tK_0(t)e^{-t^2/k_0^2r_0^2}dt .$$

Around $r = 0$, one may expand Δn_0 as

$$\Delta n = \Delta n\Big|_{r=0} + c_2r^2 . \tag{6.21}$$

Using the expansion in Eq. (6.19) we get

$$c_2 = \frac{1}{4}\left[\alpha + k_0^2 \Delta n\Big|_{r=0}\right] .$$

The nonlinear permittivity can be written as

$$\Phi = -\frac{\omega_p^2}{\omega^2}\frac{1}{n_0^0}\left[\Delta n\Big|_{r=0} + \frac{1}{4}\left(\alpha + k_0^2 \Delta n\Big|_{r=0}\right) r^2\right] . \tag{6.22}$$

6.2. Self-Focusing

In a plasma with nonlinear permittivity $\nabla \cdot \mathbf{E} = \frac{1}{\varepsilon}\mathbf{E} \cdot \nabla\varepsilon \neq 0$, in general. However, for $\nabla\varepsilon/\varepsilon \ll k$ one may ignore the $\nabla \cdot \mathbf{E}$ term in the wave equation. For waves having rapid phase variations $\widetilde{E} = \widetilde{E}_0(r, z)\exp[-i(\omega t - kz)]$ the wave equation can be written as

$$2ik\frac{\partial E_0}{\partial z} + \nabla_\perp^2 E_0 + \frac{\omega^2}{c^2}\Phi E_0 = 0 . \tag{6.23}$$

Introducing an eikonal $E_0 = A_0(r, z)e^{iS(r,z)}$ Eq. (6.23) gives[1,10]

$$k\frac{\partial A_0^2}{\partial z} + \frac{\partial S}{\partial r}\frac{\partial A_0^2}{\partial r} + A_0^2\nabla_\perp^2 S = 0 , \tag{6.24}$$

$$2k\frac{\partial S}{\partial z} + \left(\frac{\partial S}{\partial r}\right)^2 = \frac{1}{A_0}\nabla_\perp^2 A_0 + \frac{\omega^2}{c^2}\Phi = 0 . \tag{6.25}$$

Following the treatment given in Chapter 2 for an initially Gaussian beam we construct the Ansatz,

$$A_0 = \frac{A_{00}}{f(z)}e^{-r^2/2r_0^2 f^2(z)} \tag{6.26}$$

conserving power flux. Using Eq. (6.26), Eqs. (6.24) and (6.25) can be written as

$$\nabla_\perp^2 S - \frac{\partial S}{\partial r}\frac{2r}{r_0^2 f^2} - \frac{2k}{f}\left(1 - \frac{r^2}{r_0^2 f^2}\right)\frac{df}{dz} = 0 ,$$

$$2k\frac{\partial S}{\partial z} + \left(\frac{\partial S}{\partial r}\right)^2 = -\frac{1}{r_0^2 f^2}\left(2 - \frac{r^2}{r_0^2 f^2}\right) + \frac{\omega^2}{c^2}\Phi = 0 . \tag{6.27}$$

We solve these equations in the paraxial ray approximations ($r^2 \ll r_0^2 f^2$) by expanding S as

$$S = \psi(z) + \frac{\beta}{2}r^2 . \tag{6.28}$$

Equating powers of r on both sides of the above equations we obtain

$$\beta = \frac{k}{f}\frac{df}{dz}, \tag{6.29}$$

$$2k\frac{d\psi}{dz} = -\frac{2}{r_0^2 f^2} + \frac{\omega^2}{c^2}\Phi(r=0) \tag{6.30}$$

and

$$\frac{d^2 f}{dz^2} = \frac{1}{R_d^2 f^3} + \left.\frac{\partial \Phi}{dr^2}\right|_{r=0} \frac{f}{\varepsilon_0} \tag{6.31}$$

where $R_d = kr_0^2$,

$$\left.\frac{\partial \Phi}{\partial r^2}\right|_{r=0} = -\frac{A_{00}^2}{f^4 r_0^2}\left.\frac{\partial \Phi}{\partial A_0^2}\right|_{r=0}.$$

In the four cases of (i) relativistic mass nonlinearity, (ii) ponderomotive nonlinearity, (iii) ohmic heating nonlinearity without thermal conduction and (iv) ohmic heating nonlinearity with thermal conduction, $\partial\Phi/\partial r^2|_{r=0}$ can be explicitly written as

$$\left.\frac{\partial \Phi}{\partial r^2}\right|_{r=0} = -\frac{\omega_p^2 \alpha_1 A_{00}^2}{\omega^2 r_0^2 f^4 (1+\alpha_1 A_{00}^2/f^2)^{3/2}}, \tag{i}$$

$$= -\frac{\omega_p^2 \alpha_2 A_{00}^2}{\omega^2 r_0^2 f^4} e^{-\alpha_2 A_{00}^2/f^2}, \tag{ii}$$

$$= -\frac{\omega_p^2 \alpha_3 A_{00}^2}{\omega^2 r_0^2 f^4 (1+\alpha_3 A_{00}^2/f^2)^{2}}, \tag{iii}$$

$$= -\frac{\omega_p^2 \alpha_4 A_{00}^2}{4\omega^2 f^2}(1 - I_0(0)I), \tag{iv}$$

(6.32)

where

$$\alpha_1 = \frac{e^2}{m^2\omega^2 c^2}, \qquad \alpha_2 = \frac{e^2}{m^2\omega^2 v_{th}^2},$$

$$\alpha_3 = \frac{e^2}{6\delta m^2 \omega^2 v'^2_{th}}, \qquad \alpha_4 = \frac{e^2\nu^2}{2m^2\omega^2 v_{th}^4},$$

$$I = \int_0^\infty tK_0(t)e^{-t^2/k_0^2 r_0^2 f^2}\,dt.$$

Multiplying Eq. (6.31) by $2df/dz$ and integrating over z from 0 to z we obtain, for an initially plane wave front ($df/dz|_{z=0} = 0$)

$$\left(\frac{df}{dz}\right)^2 = \left(1 - \frac{1}{f^2}\right)\frac{1}{R_d^2} - P(f) + P(1) \tag{6.33}$$

where

$$P(f) = \frac{2\omega_p^2}{\omega^2 r_0^2 \varepsilon_0} \frac{1}{(1+\alpha_1 A_{00}^2/f^2)^{1/2}}\,, \tag{i}$$

$$= \frac{\omega_p^2}{\omega^2 r_0^2 \varepsilon_0} e^{-\alpha_2 A_{00}^2/f^2}\,, \tag{ii}$$

$$= \frac{\omega_p^2}{\omega^2 r_0^2 \varepsilon_0} \frac{1}{(1+\alpha_3 A_{00}^2/f^2)}\,, \tag{iii}$$

$$= \frac{\omega_p^2 \alpha_4 A_{00}^2}{4\omega^2 \varepsilon_0} \ln f\,; \tag{iv}$$

(6.34)

the last expression (iv) has been written in the limit of $k_0^2 r_0^2/4f^2 \ll 1$. At the focal point $f = f_{\min}$, $df/dz = 0$. From Eq. (6.33) we obtain a transcendental equation for $f_{\min}$

$$P(f_{\min}) - P(1) = -\left(\frac{1}{f_{\min}^2} - 1\right)\frac{1}{R_d^2}\,. \tag{6.35}$$

which can be solved graphically to obtain $f_{\min}$ as a function of power. Another integration of Eq. (6.33) is difficult analytically. However, when nonlinearity is weak Eq. (6.33) can be integrated in the first three cases to give

$$f^2 = 1 + z^2/z_f^2\,, \tag{6.36}$$

where

$$z_f^{-2} = R_n^{-2} - R_d^{-2}\,,$$

$$R_n = r_0 \frac{\omega}{\omega_p\sqrt{\alpha A^2}}\,, \tag{6.37}$$

$\alpha = \alpha_1, \alpha_2, \alpha_3$ in the three cases (i), (ii), and (iii) respectively. R_n is known as the nonlinear self-focusing length. When $R_n < R_d$, the nonlinear refraction supersedes diffraction and beam converges, f taking smaller values.

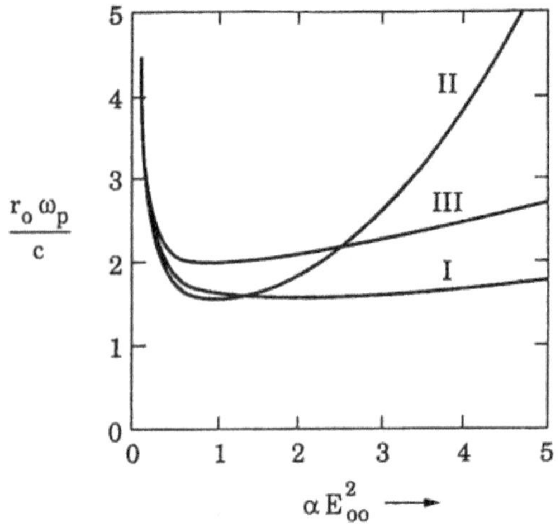

Fig. 6.2. Radius of self trapped radiation as a function of axial intensity. For relativistic nonlinearity (I) $\alpha = e^2/m^2\omega^2c^2$, ponderomotive nonlinearity (II) $\alpha = e^2/m^2\omega^2v_{th}^2$ and thermal nonlinearity without thermal conduction (III) $\alpha = e^2/6\delta m^2\omega^2 v_{th}'^2$.

However, as the beam converges, saturation effects of nonlinearity become important and f varies in a periodic manner with z. For an elaborate discussion of periodic self-focusing the reader is referred to a review article by Sodha *et al.*

The threshold for self-focusing can be obtained from Eq. (6.31), balancing diffraction and self-focusing terms,

$$\left.\frac{\partial\Phi}{\partial r^2}\right|_{\substack{r=0\\ f=1}} = -\frac{\varepsilon_0}{R_d^2} \,. \tag{6.38}$$

Equation (6.38) determines the beam radius as a function of power density for self-trapping. It provides the boundary between self-focusing and defocusing (cf. Fig. 6.2). One may note that the radius of a self-trapped radiation attains a minimum $r_0c/\omega_p \sim 2$ at $\alpha A_{00}^2 \sim 1$. At higher power densities $\alpha A_{00}^2 > 1$, the nonlinear refraction starts weakening and the spot size of a self trapped radiation becomes larger.

When $r_0, \alpha A_{00}^2$ are above the self-trapping curve, the electromagnetic beam undergoes self-focusing, attaining a spot size $r_0 f_{\min}$ at the focus. We have plotted $f_{\min}$ as a function of incident power density in Fig. 6.3.

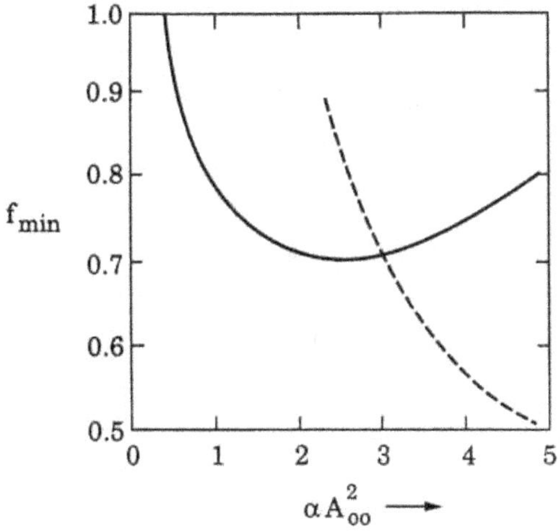

Fig. 6.3. Size of the focal spot $r_0 f_{\min}$ as a function of the power density of the beam. For thermal self-focusing with thermal conduction (- - - -) $\alpha = e^2 \nu^2 \omega_p^2 r_0^4 / 8m^2 \omega^2 v_{th}^2 c^2$. For relativistic self-focusing (- - - -) $\alpha = e^2/m^2\omega^2 c^2$, $r_0 \omega_p / c = 1.9$. The behavior of $f_{\min}$ in the other cases is similar to that of the relativistic self-focusing.

6.3. Filamentation Instability

Now we examine the stability of a uniform electromagnetic wave to a small amplitude perturbation,

$$\mathbf{E} = [\mathbf{A}_0 + A_1(x, z)] e^{-i(\omega t - kz)} \tag{6.39}$$

where $\mathbf{A}_0 \| \mathbf{A}_1$, $\nabla \ln A_1 \ll k$, $k = (\omega/c)\varepsilon_0^{1/2}$, $\varepsilon_0 = 1 - \omega_p^2/\omega^2\gamma_0$. In the case when $v \to c$ it is useful to consider a circular polarization. However, to avoid complications we limit ourselves to quadratic nonlinearity, $\alpha A_0^2 \ll 1$ and consider linear polarization of $\mathbf{A}_0 \| \hat{y}$. Taking A_0 to be real the effective permittivity of the plasma can be cast as

$$\varepsilon = \varepsilon_0 + \varepsilon_2 \mathbf{A}_0 \cdot (\mathbf{A}_1 + A_1^*) , \tag{6.40}$$

where (i) $\varepsilon_2 = \omega_p^2 \alpha_1 / 2\omega^2$, (ii) $\omega_p^2 \alpha_2 / 2\omega^2$, (iii) $2\omega_p^2 \alpha_3/\omega^2$, and (iv) $\omega_p^2 e^2 / 6m^2 v_{th}'^2 \delta' \omega^4$. Using Eqs. (6.39), (6.40) in the wave equation and using $\nabla \cdot (\varepsilon \mathbf{E}) = 0$ we obtain, on linearization,

$$2ik \frac{\partial A_1}{\partial z} + \frac{\partial^2 A_1}{\partial x^2} + \varepsilon_2 A_0^2 (A_1 + A_1^*) = 0 . \tag{6.41}$$

Expressing $A_1 = A_{1r} + iA_{1i}$ and separating real and imaginary parts,

$$2k\frac{\partial A_{1r}}{\partial z} + \frac{\partial^2 A_{1i}}{\partial x^2} = 0\,,$$

$$-2k\frac{\partial A_{1i}}{\partial z} + \frac{\partial^2 A_{1r}}{\partial x^2} + \frac{2k^2\varepsilon_2}{\varepsilon_0}A_0^2 A_{1r} = 0\,. \tag{6.42}$$

For $A_{1r}, A_{1i} \sim E^{i(q_\perp x + q_\parallel z)}$, Eqs. (6.42) straight away yield the dispersion relation

$$q_\parallel = -i\frac{q_\perp}{2k}\left[2k^2\frac{\varepsilon_2 A_0^2}{\varepsilon_0} - q_\perp^2\right]^{1/2}\,. \tag{6.43}$$

The instability occurs when

$$q_\perp^2 < k^2 2\frac{\varepsilon_2 A_0^2}{\varepsilon_0}\,.$$

Maximum growth $\gamma_{\max}$ occurs for $q_\perp = q_{opt}$,

$$q_{\rm opt}^2 = k^2\varepsilon_2 A_0^2/\varepsilon_0$$

$$\gamma_{\max} = \frac{k}{2\varepsilon_0}\varepsilon A_0^2 \tag{6.44}$$

The growth length $R_g \equiv \gamma_{\max}^{-1}$ can be written as

$$R_g = \frac{2}{q_{\rm opt}}\left(\frac{\varepsilon_0}{\varepsilon_2 A^2 y_0}\right)^{1/2}$$

which is the same as the self focusing length R_n (cf. Eq. (5.37)) of a Gaussian beam of radius $2/q_{\rm opt}$.

In most of the experiments the main beam has finite transverse extent and undergoes self-focusing. However, smaller-scale pertubation grow faster than the main beam self-focusing, hence the beam tends to break up into filament before reaching the focus. In the nonlinear state the filament attains a size $\sim c/\omega_p$. The inhomogeneity of the plasma does not suppress filamentation and self-focusing. Nevertheless it would be neccessary for the length of the underdense region to exceed the growth length or self-focusing length for these processes to be important. In the present-day laser-plasma interaction experiments one encounters large under-dense plasmas where self-focusing and filamentation are important. In wakefield and beat-wave accelerators also, self-focusing occurring through the relativistic and ponderomotive nonlinearities is important. In the ionospheric modification experiments self-focusing occurs via the ohmic nonlinearity.

References

1. M. S. Sodha, A. K. Ghatak, and V. K. Tripathi, *Prog. in Optics*, **13**, 169 (North Holland, 1976).
2. A. J. Palmer, *Phys. Fluids* **14**, 2714 (1971).
3. F. Perkins and E. Valeo, *Phys. Rev. Lett.* **32**, 1234 (1974); C. E. Max, J. Aarons and A. B. Langdon, *Phys. Rev. Lett.* **33**, 209 (1974); R. D. Jones, W. C. Mead, S. V. Coggeshall, C. H. Aldrich, J. L. Norton, G. D. Pollak, and J. M. Wallace, *Phys. Fluids* **31**, 1249 (1988).
4. G. Z. Sun, E. Ott, Y. C. Lee, and P. Guzdar, *Phys. Fluids* **30**, 526 (1987).
5. J. C. Solem, T. S. Luk, K. Boyer, and C. K. Rhodes, *IEEE J. Quant. Electron* **25**, 2423 (1989).
6. H. Hora, *Physics of Laser Driven Plasmas* (Wiley, New York, 1981).
7. P. K. Kaw, G. Schmidt, and T. Wilcox, *Phys. Fluids* **16**, 1522 (1973).
8. C. S. Liu and P. K. Kaw, *Adv. in Plasma Physics*, eds. A. Simon and W. B. Thompson **6**, 83 (1976).
9. V. L. Ginzburg, *The Propagation of Electromagnetic Waves in Plasmas* (Pergamon, 1960).
10. S. A. Akhmanov, A. P. Sukhorukov, and R. V. Khokhlov, *Sov. Phys. Usp.* **10**, 609 (1968).

CHAPTER 7

PARAMETRIC INSTABILITIES IN A HOMOGENEOUS PLASMA

A large amplitude wave in a plasma, i.e., a pump, induces electron oscillations with velocity $\mathbf{v}_0$. One may wish to ask whether the free energy contained in this oscillatory motion could drive plasma eigenmodes unstable. For example, consider a Langmuir wave in the presence of an electromagnetic pump wave. The space charge oscillations due to the Langmuir wave and the oscillatory electron velocity due to the pump constitute a nonlinear current at sum and difference frequency. When frequency and wave vector of the nonlinear current satisfy the dispersion relation of an electromagnetic wave in the plasma, this mode (sideband) is resonantly excited. The sideband and pump waves in turn exert a $\mathbf{v}_0 \times \mathbf{B}_1$ ponderomotive force on the electrons at the frequency and wave vector of the original Langmuir mode, further amplifying the Langmuir wave. This feedback mechanism may thus lead to simultaneous growth of the Langmuir wave and the electromagnetic sideband. It is an example of parametric instability of Raman scatttering in plasmas. In this process a pump wave photon of energy and momentum $(\hbar\omega_0, \hbar\mathbf{k}_0)$ produces decay wave photons $(\hbar\omega, \hbar\mathbf{k})$ and $(\hbar\omega_1, \hbar\mathbf{k}_1)$. Energy and momentum conservation demand,

$$\begin{aligned} \hbar\omega_0 &= \hbar\omega + \hbar|\omega_1|, \\ \hbar\mathbf{k}_0 &= \hbar\mathbf{k} + \operatorname{sign}(\omega_1)\hbar\mathbf{k}_1 \ . \end{aligned} \tag{7.1}$$

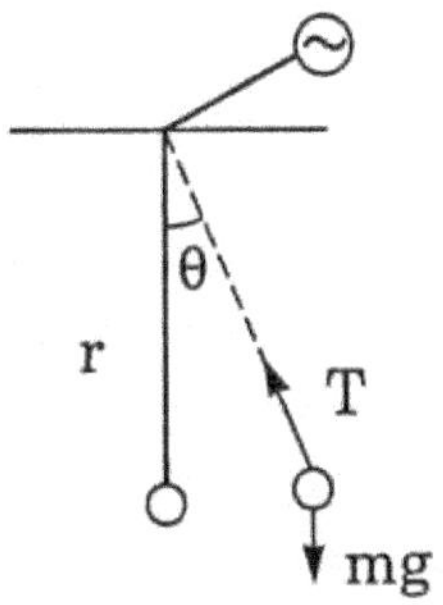

Fig. 7.1A. Simple pendulum with modulated length $r = l_0 + l_{10} \cos \omega_0 t$.

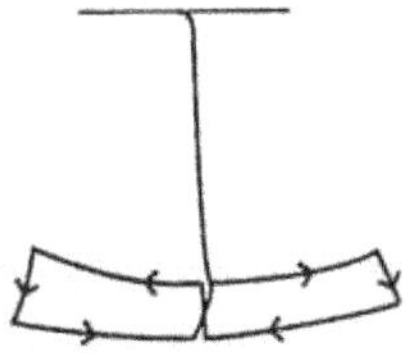

Fig. 7.1B. Time history of the movement of the center of mass of the swing and the mass, effectively modulating the length of the pendulum.

Equation (7.1) represents the phase matching conditions also for a three-wave resonant interaction. In the following analysis we assume, without any loss of generality $\omega < |\omega_1|$, and designate the low frequency mode $\omega, \mathbf{k}$ and the sideband $\omega_1, \mathbf{k}_1$.

Parametric instability conventionally implies the modulation of a parameter, e.g., plasma density or particle drift, and subsequent excitation of eigenmodes of the system. Before taking up a detailed analysis of parametric instability in a plasma we consider two simple examples.

7.1. A Harmonic Oscillator

Consider a simple pendulum of length r. Let the length be modulated periodically by pulling the string in and out of the suspension point,

$$r = l_0 + l_1; \qquad l_1 \ll l_0 , \tag{7.2}$$

where $l_1 = l_{10} \cos \omega_0 t$ and l_0 is the length without modulation (cf. Fig. 7.1). If $l_1 = 0$, the frequency of oscillation is

$$\Omega_0 = \sqrt{g/l_0}\ , \tag{7.3}$$

where g is the acceleration due to gravity. When l_1 is finite, the radial and azimuthal equations of motion of the pendulum of a mass m can be written as

$$\ddot{r} - r\dot{\theta}^2 = \frac{T}{m} - g\cos\theta \tag{7.4}$$

$$\frac{1}{r}\frac{d}{dt}(r^2\dot{\theta}) = -g\sin\theta \tag{7.5}$$

where θ is the angular displacement and T is the tension in the string. For $\theta \ll 1$ we introduce $x = r\theta$, then using Eq. (7.2), Eq. (7.5) takes the form

$$\ddot{x} + \Omega_0^2[1 + \varepsilon\cos\omega_0 t]x = 0\ , \tag{7.6}$$

where

$$\varepsilon = \left(\frac{\omega_0^2}{\Omega_0^2} - 1\right)\frac{l_{10}}{l_0}\ .$$

This is Mathieu's equation having unstable solutions in certain parameter space.[1] We solve it by perturbation theory in two cases: (i) $\omega_0 \simeq 2\Omega_0$ and (ii) $\omega_0 \simeq \Omega_0$.

Case (I) $\omega_0 \simeq 2\Omega_0$

Fourier-analysing Eq. (7.6) we obtain

$$\begin{aligned}(\omega^2 - \Omega_0^2)x_\omega - \frac{\varepsilon\Omega_0^2}{2}x_{\omega_1} &= 0\ ,\\ (\omega^2 - \Omega_0^2)x_{\omega_1} - \frac{\varepsilon\Omega_0^2}{2}x_\omega &= 0\ ,\end{aligned} \tag{7.7}$$

where $\omega_1 \equiv \omega - \omega_0$ and we have retained only those terms having frequency close to Ω_0. Equations (7.7) yield a nonlinear dispersion relation

$$(\omega^2 - \Omega_0^2)(\omega_1^2 - \Omega_0^2) = \left(\frac{\varepsilon\Omega_0^2}{2}\right)^2\ . \tag{7.8}$$

Defining a frequency mismatch $\Delta \equiv \omega_0 - 2\Omega_0$, $\omega = \Omega_0 + G$ where Δ, $G \ll \omega_0$, Eq. (7.8) yields,

$$G = \frac{1}{2}\left[\Delta \pm \sqrt{\Delta^2 - \frac{1}{4}\varepsilon^2\Omega_0^2}\right]\ . \tag{7.9}$$

For $\varepsilon\Omega_0 > 4\Delta$ unstable solutions exist with a growth rate

$$\gamma = \frac{1}{2}\sqrt{\varepsilon^2\Omega_0^2/4 - \Delta^2} \tag{7.10}$$

whereas the real frequency $\omega_r = \omega_0/2$, i.e., it is a frequency-locked oscillation, ω_r depending only on ω_0 and not on the natural frequency of oscillation. The growth rate maximizes to $\gamma_{\max} = \varepsilon\Omega_0/4$ for $\Delta = 0$.

If one incorporates linear damping in Eq. (7.6), i.e., adds a term $2\Gamma\dot{x}$ to the left side of Eq. (7.6), the nonlinear dispersion relation takes the form

$$[\omega^2 - \Omega_0^2(1 - 2i\Gamma)][\omega_1^2 - \Omega_0^2(1 + 2i\Gamma)] = \frac{\varepsilon^2\Omega_0^4}{4} \tag{7.11}$$

giving

$$\begin{aligned} G &= \frac{1}{2}\left[\Delta \pm \sqrt{\Delta^2 - \varepsilon^2\Omega_0^2/4}\right] - i\Gamma\,, \\ \gamma &= \frac{1}{2}\sqrt{\frac{\varepsilon^2\Omega_0^2}{4} - \Delta^2} - \Gamma\,. \end{aligned} \tag{7.12}$$

The threshold for the parametric instability is

$$\varepsilon^2 \geq \frac{4}{\Omega_0^2}(4\Gamma^2 + \Delta^2)\,. \tag{7.13}$$

If damping and frequency mismatch are negligible ($\Gamma = 0, \Delta = 0$) the instability can be driven by an infinitesimal modulation. One may cite an interesting example of such a parametric excitation, viz., a man riding a swing. To acquire higher amplitude he has to move up and down his center of gravity, modifying the effective length of the swing at twice the frequency of the swing.

Case (II) $\omega_0 \simeq \Omega_0$

In this case one would excite oscillations around a natural frequency ($\pm\Omega_0$). The beat of these oscillations with ω_0 produces a low frequency oscillation with $\omega \ll \omega_0$. A second harmonic is also generated but its amplitude is down by $\frac{1}{3}$ as compared to that of the ω mode, hence dropped.

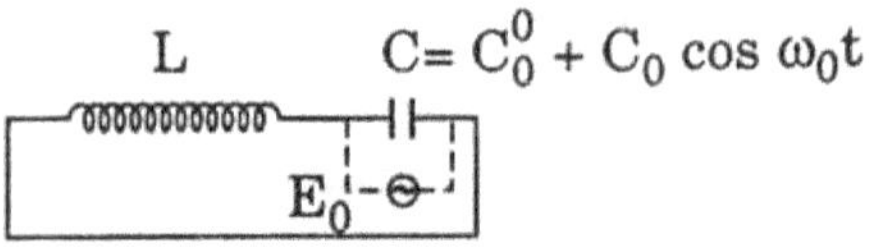

Fig. 7.2. A parametric oscillator with a nonlinear capacitor.

Fourier-analysing Eq. (7.6) we obtain

$$(\omega^2 - \Omega_0^2)x_\omega + \frac{\varepsilon\Omega_0^2}{2}(x_{\omega_1} + x_{\omega_2}) = 0,$$

$$(\omega_1^2 - \Omega_0^2)x_{\omega_1} + \frac{\varepsilon\Omega_0^2}{2}x_\omega = 0\ , \tag{7.14}$$

$$(\omega_2^2 - \Omega_0^2)x_{\omega_2} + \frac{\varepsilon\Omega_0^2}{2}x_\omega = 0$$

which yield

$$\omega^2 - \Omega_0^2 - \left[\frac{\varepsilon\Omega_0^2}{2}\right]^2 \left[\frac{1}{\omega_1^2 - \Omega_0^2} + \frac{1}{\omega_2^2 + \Omega_0^2}\right]\ , \tag{7.15}$$

where $\omega_1 = \omega - \omega_0$, $\omega_2 = \omega + \omega_0$. Defining $\Delta = \omega_0 - \Omega_0$, Δ, $\omega \ll \omega_0$, we have from Eq. (7.15).

$$\omega^2 = \Delta^2 + \frac{\varepsilon^2\Omega_0\Delta}{4}\ . \tag{7.16}$$

A purely growing solution occurs when $\Delta < 0$ and $|\Delta| < \varepsilon^2\Omega_0/8$. Maximum growth occurs when $\Delta = -\varepsilon^2\Omega_0/8$,

$$\gamma_{\text{max}} = \frac{\varepsilon^2\Omega_0}{8} \tag{7.17}$$

Inclusion of damping yields

$$\gamma = \sqrt{-\Delta\left(\frac{\varepsilon^2\Omega_0}{4} - \Delta\right)} - \Gamma\ . \tag{7.18}$$

One may note that two frequencies ω, $\pm\omega_0$ are excited simultaneously, both having the same growth rate but the amplitude of the zero frequency mode (cf. Eq. (7.14)) is $\sim \varepsilon/2$ times that of the ω_0 frequency mode.

As another example of a parametric oscillator one may consider a LC circuit (cf. Fig. 7.2) in which capacitance is a nonlinear element, i.e., by the application of an ac electric field the capacitance has been modulated

$C = C_0^0 + C_0 \cos\omega_0 t$. Using the Mesh theorem, the equation governing the current flow can be written as

$$\frac{d^2 I}{dt^2} + \Omega_0^2 \left(1 - \frac{C_0}{C_0^0} \cos\omega_0 t\right) I = 0 \ , \tag{7.19}$$

where $\Omega_0 = 1/\sqrt{LC_0^0}$. This equation is the same as Eq. (7.6).

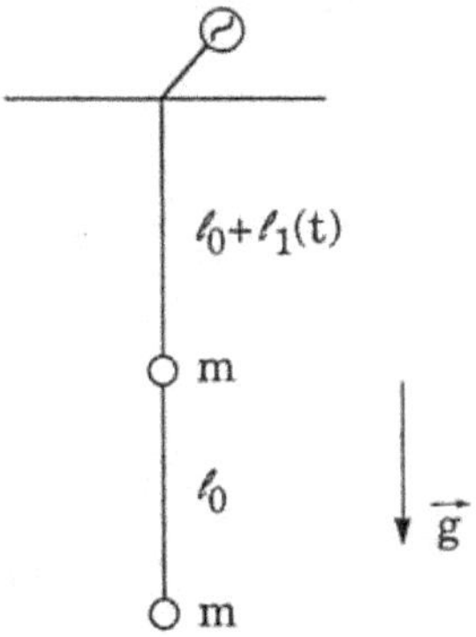

Fig. 7.3. Two masses suspended through two strings and constrained to execute only vertical motion form a system with two modes of oscillation. Modulation of length of one of the strings at the sum of the frequencies of natural oscillation leads to the parametric excitation of both the modes.

7.2. Parametric Oscillator with Two Degrees of Freedom

A pump can simultaneously excite two eigenmodes of a system with two degrees of freedom. Consider a model example (cf. Fig. 7.3) in which a mass m is suspended by an elastic string of length l. Another mass m is suspended to the first one by another string of length l_0 and the system is free to oscillate only along the vertical. Let the strings be characterized by an area of cross-section A and Young's modulus Y. Also let

$$l = l_0 = l_1(t) \ , \tag{7.20}$$

where $l_1 = l_{10} \cos\omega_0 t$, $l_1 \ll l_0$. The motions of the two masses are governed by

$$m\ddot{x}_1 = -YA\frac{x_1 - l}{l} + YA\frac{x_2 - x_1 - l_0}{l_0} + mg\ ,$$

$$m\ddot{x}_2 = -Ya\frac{x_2 - x_1 - l_0}{l_0} + mg \tag{7.21}$$

where x_1 and x_2 are the position of the two masses from the ceiling. Defining $\Omega_0^2 = YA/ml_0$, Eqs. (7.21) can be written as

$$\ddot{x}_1 = -\Omega_0^2\left[\left(2 - \frac{l_1}{l_0}\right)x_1 - x_2\right] + g,$$

$$\ddot{x}_2 = -\Omega_0^2[x_2 - x_1 - l_0] + g\ . \tag{7.22}$$

Expanding x_1, x_2 in time independent and time dependent parts

$$x_1 = x_{10} + y_1(t), \qquad x_2 = x_{20} + y_2(t)\ ,$$

we get

$$\ddot{y} = -\Omega_0^2\left[\left(2 - \frac{l_1}{l_0}\right)y_1 - y_2\right]\ , \tag{7.23}$$

$$\ddot{y}_2 = -\Omega_0^2(y_2 - y_1)\ . \tag{7.24}$$

Multiply Eq. (7.24) by α and add it to Eq. (7.23),

$$\frac{d^2}{dt^2}(y_1 + \alpha y_2) = -\Omega_0^2(2 - \alpha)\left[y_1 + \frac{\alpha - 1}{2 - \alpha}y_2\right] + \Omega_0^2\frac{l_1}{l_0}y_1\ , \tag{7.25}$$

choosing α such that $\alpha = (\alpha - 1)/(2 - \alpha)$ we get two independent modes of oscillation when $l_1 = 0$, viz., $\alpha = \alpha_1, \alpha_2$;

$$\alpha_{1,2} = \frac{1 \pm \sqrt{5}}{2}\ . \tag{7.26}$$

Corresponding eigen frequencies are $\Omega_{1,2} = \frac{\Omega_0}{\sqrt{2}}[3 \mp \sqrt{5}]^{1/2}$. Defining $\psi_1 = y_1 + \alpha_1 y_2$, $\psi_2 = y_1 + \alpha_2 y_2$ the two coupled mode equations for ψ_1 and ψ_2 (for $l_1 \neq 0$) can be written as

$$\frac{d^2\psi_1}{dt^2} + \Omega_1^2\psi_1 = \Omega_0^2\frac{l_1}{l_0}\frac{\alpha_2\psi_1 - \alpha_1\psi_2}{\alpha_2 - \alpha_1}, \tag{7.27}$$

$$\frac{d^2\psi_2}{dt^2} + \Omega_2^2\psi_2 = \Omega_0^2\frac{l_1}{l_0}\frac{\alpha_2\psi_1 - \alpha_1\psi_2}{\alpha_2 - \alpha_1}\ . \tag{7.28}$$

We solve this set for $\omega_0 \simeq \Omega_1 + \Omega_2$. Fourier-transforming Eqs. (7.27) and (7.28) we get

$$(\omega^2 - \Omega_1^2)\psi_1(\omega) = \frac{\Omega_0^2 l_1}{2l_0} \frac{\alpha_1 \psi_2(\omega_1)}{\alpha_2 - \alpha_1} ,$$
$$(\omega_1^2 - \Omega_1^2)\psi_2(\omega_1) = -\frac{\Omega_0^2 l_1}{2l_0} \frac{\alpha_2 \psi_1(\omega)}{\alpha_2 - \alpha_1} , \tag{7.29}$$

giving the nonlinear dispersion relation

$$(\omega^2 - \Omega_1^2)(\omega_1^2 - \Omega_2^2) = \frac{\Omega_0^4 l_{10}^2}{20 l_0^2} , \tag{7.30}$$

where $\omega_1 = \omega - \omega_0$. Expanding ω, ω_1 around Ω_1 and $-\Omega_2$ respectively, $\omega = \Omega_1 + i\gamma$, $\omega_1 = -\Omega_2 + i\gamma$ we obtain the growth rate

$$\gamma = \Omega_0 \frac{l_{10}}{4\sqrt{5} l_0} . \tag{7.31}$$

Thus a parametric oscillator can be operated to excite two eigenmodes at the expense of the energy of the pump.

7.3. Parametric Coupling in a Plasma

An unmagnetized plasma supports three modes of wave propagation, viz., Langmuir ($\omega^2 = \omega_p^2 + 3k^2 v_{th}^2/2$), ion acoustic ($\omega \sim kc_s$) and electromagnetic ($\omega^2 = \omega_p^2 + k^2c^2$) modes. Nonlinear coupling of these modes has been a subject of great fascination and importance for laser plasma interaction, since it was first pointed out by one of the author (CSL) in 1972. Here we consider the pump to be either an electromagnetic wave or a linearly mode converted Langmuir wave and examine the initial stage of parametric instability. For an electromagnetic pump the possible channels of resonant decay are[2–10] (Fig. 7.4):

(I) Stimulated Raman Scattering (SRS) producing a pair of Langmuir and electromagnetic waves at densities below quarter critical $n_0^0 \leq n_{cr}/4$ (n_{cr} is the critical density where the plasma frequency is equal to the pump frequency).

(II) Stimulated Brillouin Scattering (SBS) producing an ion acoustic wave and an electromagnetic wave sideband at $n_0^0 \leq n_{cr}$

(III) Decay instability producing an ion acoustic wave and a Langmuir wave sideband near the critical layer.

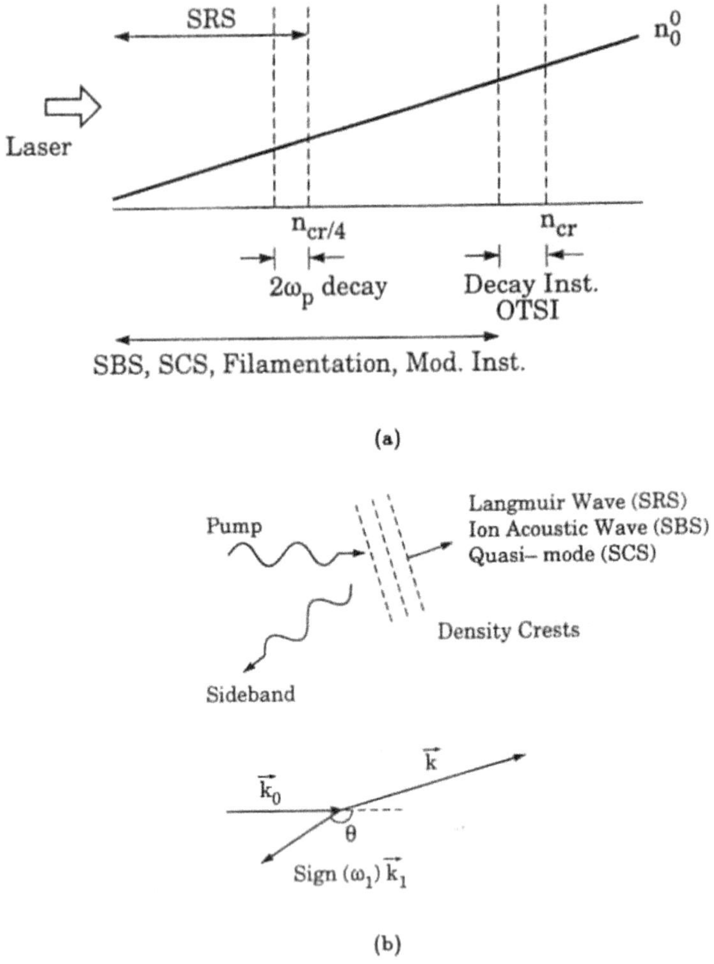

Fig. 7.4. (a) Location of various parametric processes in an inhomogeneous plasma. (b) Schematic of scattering process.

(IV) Two-plasmon decay, producing two Langmuir waves near quarter critical density $n_0^0 \sim n_{cr}/4$.

Decay into two electromagnetic waves is not allowed kinematically because the pump wave number is always greater than the sum of the wave numbers of the decay waves $|k_0| > |k| + |k_1|$, hence the **k** matching condition is not satisfied. In processes (I) and (III) the density perturbation association with the low frequency mode and the oscillatory electron velocity due to the pump constitute a nonlinear current driving the sideband. The pump and the sideband exert a low frequency ponderomotive force on the electrons driving the density perturbation. In channel IV both the decay waves are Langmuir waves having nearly equal frequencies, hence both types of nonlinearities are important for both the waves.

There exists another possibility. The low frequency perturbation, instead of being a resonant eigenmode, could be strongly Landau damped on electrons and ions. In the presence of the pump, such a perturbation, in conjunction with the oscillatory velocity of electrons (due to the pump), produces a nonlinear current producing an electromagnetic sideband. The sideband and the pump exert a ponderomotive force on the electrons to drive the perturbation (a quasimode) as a beat wave. The sideband and quasi-mode grow with time while strongly heating the particles. This process is called collective or stimulated Compton scattering, sometimes termed nonlinear Landau damping. It can be viewed as the decay of a pump photon into a lower frequency photon while the balance of energy and momentum is shared by the particles:

$$\begin{aligned} \hbar\omega_0 &= \hbar|\omega_1| + \Delta\varepsilon \\ \hbar\mathbf{k}_0 &= \mathrm{sign}\,(\omega_1)\hbar\mathbf{k}_1 + \Delta\mathbf{p} \end{aligned} \tag{7.32}$$

where $(\omega_0, \mathbf{k}_0)$ and $(\omega_1, \mathbf{k}_1)$ refer to the pump and the daughter wave (sideband) and $\Delta\varepsilon$ and $\Delta\mathbf{p}$ are the energy and momentum imparted to the particle. For a nonrelativistic particle of mass m and initial velocity $\mathbf{v}$, $\Delta\varepsilon \simeq m\mathbf{v}\cdot\Delta\mathbf{v}$, $\Delta\mathbf{p} = m\Delta\mathbf{v}$ where $\Delta\mathbf{v}$ is the change in the velocity in the process. Equations (7.32) yield (for $\omega_1 > 0$),

$$\omega_0 - \omega_1 = (\mathbf{k}_0 - \mathbf{k}_1)\cdot\mathbf{v}\ . \tag{7.33}$$

For a Langmuir wave pump the decay channels are: (I) Resonant decay into ion acoustic and Langmuir waves, producing longer-wavelength modes and (II) Nonlinear Landau damping on ions. Decay channels involving electromagnetic sidebands are relatively unimportant because they suffer from larger convection losses.

In our discussion so far we have ignored the upper sideband $\omega + \omega_0$, $\mathbf{k}+\mathbf{k}_0$, which is justified as long as it is off-resonant. Consider, for example, the coupling of a pump $(\omega_0, \mathbf{k}_0)$ with a space charge wave $\omega, \mathbf{k}$ and two electromagnetic sidebands $(\omega \mp \omega_0, \mathbf{k} \mp \mathbf{k}_0)$. If we choose the lower sideband $(\omega - \omega_0, \mathbf{k} - \mathbf{k}_0)$ to be resonant, i.e.,

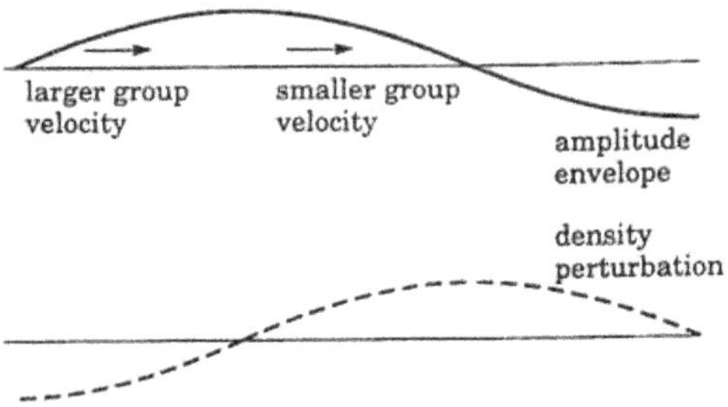

Fig. 7.5. The dynamics of modulational instability. The density perturbation is $\pi/2$ out of phase with the amplitude envelope. The change in group velocity induced by the density perturbation causes the enhancement of amplitude modulation.

$$\omega_0 - \omega = [\omega_p^2 + (\mathbf{k} - \mathbf{k}_0)^2 c^2]^{1/2} \tag{7.34}$$

then the frequency mismatch for the upper sideband is

$$\begin{aligned}\Delta\omega &\equiv \omega_0 + \omega - [\omega_p^2 + (\mathbf{k} + \mathbf{k}_0)^2 c^2]^{1/2} \\ &= 2\omega_0 - [\omega_p^2 + (\mathbf{k} - \mathbf{k}_0)^2 c^2]^{1/2} - [\omega_p^2 + (\mathbf{k} + \mathbf{k}_0)^2 c^2]^{1/2} .\end{aligned} \tag{7.35}$$

As long as $\Delta\omega$ is greater than the growth rate one can ignore the upper sideband. The possibility of an upper sideband being resonant and lower sideband being off-resonant is discarded here as it is not kinematically allowed in a plasma without dc drifts. A photon cannot produce a photon of higher energy unless the low frequency mode is a negative energy mode. Nevertheless, there are processes where both the sidebands are important:

(i) *Modulational instability*

When $\omega/k = \partial\omega_0/\partial k_0$ and $\mathbf{k}||\mathbf{k}_0$, the perturbation (not a normal mode) propagates with a phase velocity equal to the group velocity of the pump, leading to amplitude modulation of the pump wave. A small modulation in the amplitude of the pump wave exerts a ponderomotive force $\mathbf{F}_p||\mathbf{k}_0$ on the electrons. Usually the group velocity v_{go} of amplitude propagation is larger than the sound speed c_s (at which ambipolar diffusion occurs), hence, the density perturbation induced by the ponderomotive force is $\pi/2$ out of phase with it (Fig. 7.5). The modified density modulates the group velocity of the pump(v_{go} being larger where density is smaller) leading to the build up of amplitude modulation. However, for an electromagnetic pump wave ($v_{go} \gg c_s$) this process is quite weak. It is important only for a Langmuir pump.

(ii) *Filamentation instability*

When the wave vector of density perturbation $\mathbf{k} \perp \mathbf{k}_0$ and $k \ll k_0$ the density perturbation tends to concentrate (converge) electromagnetic energy around density minima where the index of refraction has maxima. The enhanced wave intensity causes more density depression leading to a purely growing perturbation and breaking of the pump wave into filaments (Fig. 7.6).

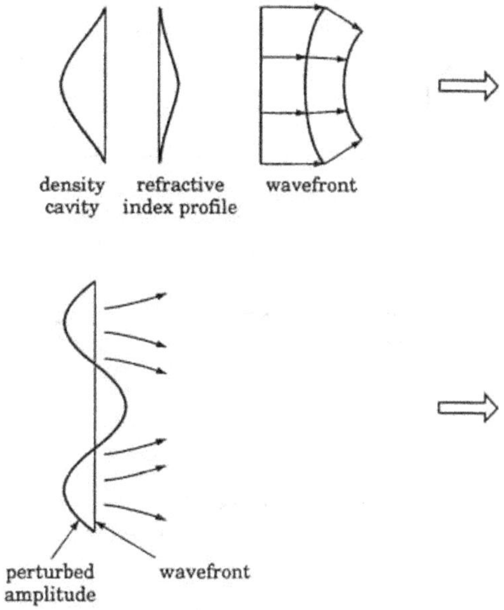

Fig. 7.6. Dynamics of filamentation. A perturbation in electromagnetic wave amplitude causes local depressions in density. Nonlinear refraction attracts more electromagnetic wave energy into the density depressions leading to the growth of the perturbation.

(iii) *Oscillating two stream instability*

In this process a long wavelength pump wave, near the critical layer, excites a short wavelength standing Langmuir wave and a purely growing density perturbation. The regions where the electron field of the pump and Langmuir waves are parallel (Fig. 7.7) the plasma is pushed away to the regions where the fields are antiparallel. The depressed density regions attract electric field energy from the neighborhood leading to deeper density depressions and enhancement of the short wavelength Langmuir mode.

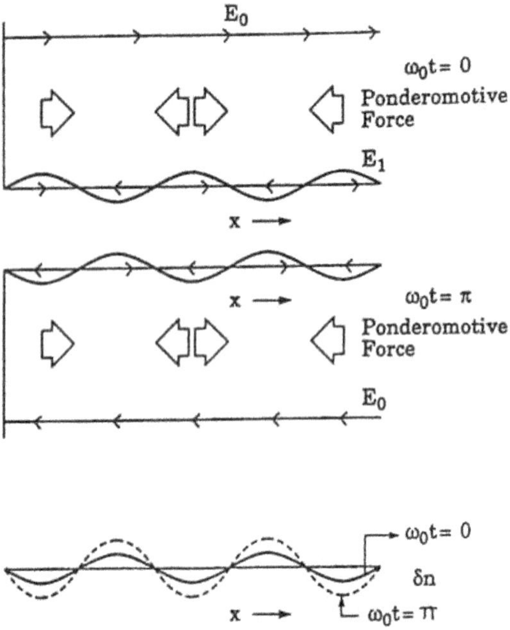

Fig. 7.7. Growth of OTSI.

In the following we follow a fluid approach to study the nonlinear wave-wave coupling in a homogeneous isotropic plasma.

7.4. Nonlinear Dispersion Relation

Consider the propagation of an electromagnetic pump wave in a plasma:

$$\mathbf{E}_0 = \mathbf{E}_0 e^{-i(\omega_0 t - k_0 z)} \,, \qquad \mathbf{B}_0 = c\mathbf{k}_0 \times \mathbf{E}_0/\omega_0 \tag{7.36}$$

It produces an oscillatory electron velocity $\mathbf{v}_0 = e\mathbf{E}_0/mi\omega_0$ (where $-e$ and m are electronic charge and mass) and excites a pair of waves, a low frequency electrostatic wave with scalar potential

$$\phi = \phi e^{-i(\omega t - \mathbf{k}\cdot\mathbf{x})} \tag{7.37}$$

and a sideband electromagnetic wave with electric and magnetic fields

$$\mathbf{E}_1 = \mathbf{E}_1 e^{-i(\omega_1 t - \mathbf{k}_1 \cdot \mathbf{x})} ,$$
$$\mathbf{B}_1 = c\mathbf{k}_1 \times \mathbf{E}_1/\omega_1 , \tag{7.38}$$

where $\mathbf{k}_1 = \mathbf{k} - \mathbf{k}_0$ and $\omega_1 = \omega - \omega_0$. The sideband is a resonant mode satisfying the linear dispersion relation: $\omega_1^2 = \omega_p^2 + k_1^2 c^2$. The linear response of electrons to the sideband is

$$\mathbf{v}_1 = eE_1/mi\omega_1 . \tag{7.39}$$

The pump and sideband waves exert a low frequency ponderomotive force on the electrons,

$$\mathbf{F}_p \equiv e\nabla\phi_p = -\frac{m}{2}[\mathbf{v}_0 \cdot \nabla\mathbf{v}_1 + \mathbf{v}_1 \cdot \nabla\mathbf{v}_0] - \frac{e}{2c}(\mathbf{v}_0 \times \mathbf{B}_1 + \mathbf{v}_1 \times \mathbf{B}_0) . \tag{7.40}$$

Solving Eq. (7.40) the ponderomotive potential turns out to be

$$\phi_p = e\mathbf{E}_0 \cdot \mathbf{E}_1/2m\omega_0\omega_1 . \tag{7.41}$$

In writing Eq. (7.40) we have used the identity $\mathrm{Re}\,\mathbf{A} \cdot \mathrm{Re}\,\mathbf{B} \equiv \frac{1}{2}\,\mathrm{Re}[\mathbf{A} \cdot \mathbf{B} + \mathbf{A} \cdot \mathbf{B}^*]$ where Re stands for "real part of". The ponderomotive and self consistent low frequency force $e\nabla(\phi + \phi_p)$ on the electrons drives density oscillations n. Using Eq. (2.57)

$$n = \frac{k^2}{4\pi e}\chi_e(\phi + \phi_p) . \tag{7.42}$$

The ponderomotive force on the ions is down by the electron to ion mass ratio, hence, their density perturbation can be taken to be linear, $n_i = -k^2\chi_i\phi/4\pi e$, where we have assumed ions to have single charge. Using n and n_i in the Poisson equation, $\nabla^2\phi = 4\pi e(n - n_i)$, we get

$$\varepsilon\phi = -\chi_e\phi_p . \tag{7.43}$$

The current density at the sideband frequency can be written as

$$\mathbf{J}_1 = -n_0^0 e\mathbf{v}_1 - \frac{1}{2}ne\mathbf{v}_0^* = -\frac{n_0^0 e^2}{mi\omega_1}\mathbf{E}_1 - \frac{k^2(1+\chi_i)}{4\pi e}\frac{e^2\mathbf{E}_0^*\phi}{2mi\omega_0} , \tag{7.44}$$

where we have used $\phi + \phi_p = -(1 + \chi_i)\phi/\xi_e$. Here n_0^0 is the equilibrium electron density. Using Eq. (7.44) in the wave equation (2.3) we get

$$D_1\mathbf{E}_1 + c^2\mathbf{k}_1\mathbf{k}_1 \cdot \mathbf{E}_1 = \frac{k^2(1+\chi_i)\omega_1}{2m\omega_0}e\mathbf{E}_0^*\phi , \tag{7.45}$$

where $D_1 = \omega_1^2 - \omega_p^2 - k_1^2c^2$. Taking the scalar product of Eq. (7.45) with $\mathbf{k}_1$ one obtains $\mathbf{k}_1 \cdot \mathbf{E}_1$ which when used back in Eq. (7.45) yields

$$D_1 E_1 = \frac{k^2\omega_1 e(1+\chi_i)}{2m\omega_0}\left(\mathbf{E}_0^* - \frac{\mathbf{k}_1}{k_1^2}\mathbf{k}_1 \cdot \mathbf{E}_0^*\right) . \tag{7.46}$$

Equations (7.43) and (7.46), on using the expression for ϕ_p field nonlinear dispersion relation, become

$$\varepsilon D_1 = -\chi_e(1+\chi_i)\frac{k^2|v_0|^2}{4}(1-\cos^2\delta_1) , \tag{7.47}$$

where δ_1 is the angle between $\mathbf{E}_0$ and $\mathbf{k}_1$. Equation (7.47) is valid for stimulated Raman, Compton and Brillouin scattering processes. If one employs kinetic theory one recovers the same dispersion relation. The kinetic effects are contained in electron and ion susceptibilities. The sideband wave is not affected by kinetic effects. However, it may suffer damping due to collisions. In this case D_1 is modified to

$$D_1 = \omega_1^2 - \omega_p^2\left(1 - i\frac{\nu}{\omega_1}\right) - k_1^2c^2 . \tag{7.48}$$

Equation (7.47) in the absence of the pump wave ($v_0 \to 0$) yields $D_1 = 0$ and/or $\varepsilon = 0$. In a resonant decay process we look for a solution of Eq. (7.47) near the simultaneous zeros of ε and D_1. Let $\omega = \omega_r - i\Gamma$ and $\omega_1 = \omega_{1r} - i\Gamma_1$ be the roots of $\varepsilon = 0$, $D_1 = 0$ i.e., $\varepsilon_r(\omega_r, \mathbf{k}) = 0$, $\Gamma = \varepsilon_i/(\partial\varepsilon_r/\partial\omega_r)$, $D_{1r}(\omega_{1r}, \mathbf{k}_1) = 0$. $\Gamma_1 = D_{1i}/(\partial D_{1r}/\partial\omega_{1r})$. To solve Eq. (7.47) we express $\omega = \omega_r + i\gamma$, $\omega_1 = \omega_{1r} + i\gamma$ (where $\omega_{1r} \equiv \omega_r - \omega_0$). Then

$$\begin{aligned} \varepsilon &= i(\gamma+\Gamma)\frac{\partial\varepsilon_r}{\partial\omega_r} \\ D_1 &= i(\gamma+\Gamma_1)\frac{\partial D_{1r}}{\partial\omega_{1r}} \\ (\gamma+\Gamma)(\gamma+\Gamma_1) &= \gamma_M^2 \equiv \frac{\chi_e^2 k^2|v_0|^2(1-\cos^2\delta_1)}{8(\omega_0-\omega)\partial\varepsilon/\partial\omega} , \end{aligned} \tag{7.49}$$

where subscript r has been dropped in favour of brevity. Setting growth rate $\gamma = 0$ one obtains the threshold for the parametric instability

$$\gamma_M^2 = \Gamma\Gamma_1 . \tag{7.50}$$

Above the threshold γ goes as v_0^2 at modest pump powers (when γ_M lies between Γ and Γ_1) whereas at large powers (for which $\gamma_M > \Gamma, \Gamma_1$) it goes as v_0.

In a nonlinear Landau damping process ($\omega \simeq k \cdot v_{th}$) ε has a large imaginary part, though D_1 is still nearly zero. Writing ω into real and imaginary parts $\omega = \omega_r + i\gamma$ and Taylor-expanding D_1 we get

$$\gamma = \frac{k^2|v_0|^2(1-\cos^2\delta_1)}{8(\omega_0-\omega)} \operatorname{Im}\left[\frac{\chi_e(1+\chi_i)}{\varepsilon}\right] - \Gamma_1 \,. \tag{7.51}$$

Above the threshold γ goes as v_0^2.

7.5. Stimulated Raman Scattering

In this process the pump electromagnetic wave is scattered off a Langmuir mode. Since the frequencies of the Langmuir wave and scattered sideband electromagnetic wave are each greater than ω_p, SRS occurs when $\omega_p < \omega_0/2$ i.e., below the quarter critical density. The wave number of the Langmuir wave is given by $k^2 = k_0^2 + k_1^2 - 2k_0|k_1|\cos\Theta$ where Θ is the scattering angle, the angle between $\mathbf{k}_0$ and $-\mathbf{k}_1$; k is largest for back scattering ($\Theta = \pi$). Since $|\omega_1| \simeq \omega_0 - \omega_p$, $|k_1| \simeq k_0$ in the far underdense region and falls to zero at the quarter critical layer where $|\omega_1| = \omega_p$. In the far underdense region $k \approx 2\frac{\omega_0}{c}\cos\Theta/2$. To ensure weak damping of the Langmuir wave $kv_{th}/\omega_p \leq 0.3$, the region of stimulated Raman scattering is limited to

$$\frac{n_0^0}{n_{c_r}} \geq 40\frac{v_{th}^2}{c^2}\cos^2\Theta/2 \,, \tag{7.52}$$

where $v_{th} = (2T_e/m)^{1/2}$. At lower density SRS goes over to stimulated Compton scattering. From Eq. (7.49) the growth rate for SRS, taking $\chi_e \simeq -1$, $\partial\varepsilon/\partial\omega \simeq 2/\omega$, can be written as

$$\gamma_M = \frac{k|v_0|}{4}\left[\frac{\omega(1-\cos^2\delta_1)}{\omega_0-\omega}\right]^{1/2} \tag{7.53}$$

The growth rate is largest when $\delta_1 = \pi/2$ and $k = 2k_0$ i.e., $\mathbf{E}_1 || \mathbf{E}_0 \perp \mathbf{k}_1$ and $\Theta = \pi$ (back scatter). In the underdense region

$$\gamma_M = \frac{|v_0|}{2c}\left(\frac{n_0^0}{n_{c_r}}\right)^{1/4}\omega_0 \tag{7.54}$$

for back scatter. The SRS is stabilized when the product $\Gamma\Gamma_1$ of the linear damping rates of the decay waves exceeds γ_M^2. This stabilization is quite effective in high z plasmas irradiated with short wavelength laser light. If we assume collisional damping the threshold condition becomes

$$\gamma_M \geq \frac{\nu_{ei}}{2} \frac{\omega_p}{|\omega_1|}$$

where ν_{ei} is the electron-ion collision frequency. Several experiments have demonstrated the collisional suppression of SRS.[11,12] The Raman reflectivity drops abruptly when $\nu_{ei} \geq 2\gamma_M$. For Au targets irradiated with 0.26 μm laser light collisional stabilization occurs for intensities less than about 10^{15} w/cm^2. In cases where low frequency density fluctuations exist in the plasma, the effective collision frequency (resustivity) tex is enhanced (as seen in Chapter 1) and threshold power for SRS is increased.

Coherent and incoherent ion acoustic waves are generated in processes like stimulated Brillouin scattering, plasma wave collapse etc. Particle simulations have shown that ion waves weaken Raman scattering.[13] The heated electron temperatures associated with SRS are also reduced. Fluid calculations using Zakharov equations extended to include SRS and SBS have shown strong suppression of SRS. In a one-dimensional simulation a uniform plasma slab is irradiated with an intense laser radiation. The plasma density is 0.045 n_{cr} and the plasma length is 15 λ_0. The intensity I of the incident wave rises linearly to a value $I\lambda_0^2 \simeq 1.6 \times 10^5$ Wμ/cm^2 in a time of 500 ps. The Raman-generated plasma wave grows, then decays as ion fluctuations associated with plasma wave collapse set in. Subsequently, Brillouin scattering takes over.

These simulations are in good agreement with experiments[14] in which a preformed plasma was irradiated by a CO_2 laser (10.6 μ). Langmuir and ion acoustic waves due to SRS and SBS were measured using time- and frequency-resolved Thompson scattering of a 0.53 μ probe beam. The signal at $\Delta\lambda \simeq 0$ corresponds to ion waves driven by SBS and the one at $\Delta\lambda \simeq 35$ Å to Langmuir waves driven by SRS. Langmuir waves appear early in the laser pulse and disappear as the ion wave level rises. In subsequent experiments the SRS signal was totally quenched by using a small counter streaming laser beam to enhance the SBS. Recent experiments with 0.35 μ laser light also indicate competition between SRS and SBS in denser and hotter plasmas with $T_e \sim$ 1–2keV. In particular in CH foil targets low Raman reflectivity from the plasma with $n_0^0 \sim n_{c_r}/4$ is found to be anti-correlated with the SBS signal. A probable explanation of SRS suppression by a coherent ion acoustic wave, generated through SBS, is as follows. The density troughs of the ion acoustic wave localize the Langmuir wave, severely restricting the interaction region and thus suppress the Raman reflectively. Alternatively, the ion acoustic wave couples with the Langmuir wave and the Raman scattered electromagnetic driving strongly Landau damped beat waves, incurring power loss on the SRS decay waves and suppressing the Raman instability.

7.6. Stimulated Compton Scattering

When $\omega \simeq k \cdot v_{\text{th}}$ the low frequency mode is strongly Landau damped on electrons. It is no longer an eigenmode but a driven oscillation. Since k is less than $2k_0$, the frequency of this driven oscillation (called a quasimode) $\omega \leq 2\omega_0 v_{\text{th}}/c \ll \omega_0$. Of course ω can take values close to Ω_{p} or far from it. Expressing $\chi_e = \varepsilon - 1 - \chi_i$, the growth rate (for $\Gamma_1 \simeq 0$, $\delta_1 \simeq \pi/2$) can be written as

$$\gamma = \frac{k^2|v_0|^2(1+\chi_{\text{i}})^2}{8(\omega_0 - \omega)} \frac{\chi_{\text{ei}}}{(1+\chi_{\text{i}}+\chi_{\text{er}})^2 + \chi_{\text{er}}^2} \tag{7.55}$$

where χ_{er}, χ_{ei} are the real and imaginary parts of electron susceptibility. For $\omega \simeq k v_{\text{th}} > \omega_{\text{pi}}$,

$$\gamma \approx \frac{|v_0|^2}{8\omega_0 v_{\text{th}}^2} \frac{\omega_{\text{p}}^2}{1 + \frac{\omega_{\text{p}}^2}{\omega^2} + 2\frac{\omega_{\text{p}}^4}{\omega^4}} \tag{7.56}$$

7.7. Stimulated Brillouin Scattering

In a non-isothermal plasma with $T_{\text{e}} \gg T_{\text{i}}$ the ion acoustic waves are weakly damped. The pump electromagnetic wave can undergo stimulated scattering off this mode. Since $\omega < \omega_{\text{pi}}$, $|\omega_1| \approx \omega_0$, $|k_1| \approx k_0$, $k \simeq 2k_0 \cos\Theta/2$. Taking $\omega \simeq kc_s$, $\partial\varepsilon/\partial\omega \simeq 2\omega_{\text{pi}}^2/k^2c_s^2\omega$, $\chi_e \simeq \omega_{\text{pi}}^2/k^2c_s^2$ the growth rate can be written as

$$\begin{aligned}
\gamma_{\text{M}} &\simeq \omega_{\text{pi}} \frac{|v_0|}{4c_s}\left(\frac{c_s}{c}\right)^{1/2}(1-\cos^2\delta_1)^{1/2}(2\cos\Theta/2)^{1/2} \\
&= \omega_{\text{pi}} \frac{|v_0|}{4c_s}\left(\frac{2c_s}{c}\right)^{1/2}
\end{aligned} \tag{7.57}$$

for backscatter. This expression is valid when $\gamma < \omega$. At large pump powers when $\gamma \sim \omega$, $\varepsilon(\omega, \mathbf{k})$ is no longer close to zero and the low frequency mode goes over to a reactive quasimode. In such a case Eq. (7.47) with $\varepsilon \approx (\omega^2 - k^2c_s^2)\omega_{\text{pi}}^2/k^2c_s^2\omega^2$ takes the form of a cubic

$$(\omega^2 - k^2c_s^2)\left(\omega - \frac{(k_0^2 - k_1^2)}{2\omega_0}c^2\right) = -\frac{\omega_{\rm pi}^2}{8\omega_0}k^2|v_0|^2 \tag{7.58}$$

for $\delta_1 = \pi/2$. Equation (7.58) can be solved numerically. For $\omega^2 \gg k^2c_s^2$, $\omega \gg (k_0^2 - k_1^2)c^2/2\omega_0$ it yields

$$\omega = \left[\frac{\omega_{\rm pi}^2}{8\omega_0}k^2|v_0|^2\right]^{1/3} \frac{1+i\sqrt{3}}{2} \,. \tag{7.59}$$

7.8. Nonlinear Ion Landau Damping

The electromagnetic pump wave can undergo stimulated Compton scattering off ions. For ions moving at a speed $\sim v_{\rm thi}$ to be in resonance with the beat wave $\omega, \mathbf{k}$, one must have $\omega \simeq kv_{\rm thi} \ll kv_{\rm th}$. In that case when $\chi_e \sim \omega_{\rm pi}^2/k^2c_s^2$, $k_1 \simeq k_0$, $k \simeq 2k_0\cos\Theta/2$ in Eq. (7.51), for $\Gamma_1 = 0$, $\delta_1 = \pi/2$, $\omega < \omega_{\rm pi}$ gives

$$\gamma \equiv \frac{|v_0|^2\omega_{\rm p}^2}{8v_{\rm th}^2\omega_0}\frac{T_e/T_{\rm i}}{\left(1+0.3\frac{T_e}{T_{\rm i}}\right)^2 + 0.3\frac{T_e^2}{T_{\rm i}^2}} \,. \tag{7.60}$$

The growth rate of this process is comparable to or more than that of the Compton scattering off electrons but the real frequency is smaller. It occurs in the large underdense region.

7.9. Modulational Instability

This is a process by which a pump wave gets amplitude modulated. Since an amplitude modulated wave is expressible as superposition of a carrier wave (pump) and two sidebands, we must include the upper sideband in our analysis. Let the upper sideband be $\mathbf{E}_2 = \mathbf{E}_2\exp[-i(\omega_2 t - \mathbf{k}_2\cdot\mathbf{x})]$, $\mathbf{B}_2 = c\mathbf{k}_2 \times \mathbf{E}_2/\omega_2$, with $\omega_2 = \omega + \omega_0$, $\mathbf{k}_2 = \mathbf{k} + \mathbf{k}_0$. Following the treatment of Sec. 7.4 one may write

$$\phi_p = \frac{e\mathbf{E}_0\cdot\mathbf{E}_1}{2m\omega_0\omega_1} - \frac{e\mathbf{E}_0^*\cdot\mathbf{E}_2}{2m\omega_0\omega_2} \,. \tag{7.61}$$

The resulting dispersion relation is

$$\varepsilon = -\chi_e(1+\chi_i)\frac{k^2|v_0|^2}{4}\left(\frac{1}{D_1} + \frac{1}{D_2}\right) \,, \tag{7.62}$$

where $D_2 = \omega_2^2 - \omega_{\rm p}^2 - k_2^2c^2$ and $\mathbf{k}\cdot\mathbf{E}_0$ is taken to zero. For modulational instability $\mathbf{k}||\mathbf{k}_0$, $\omega/k \approx \partial\omega_0/\partial k_0$, i.e.,

$$\omega \approx kv_{go} = kc(1-\omega_p^2/\omega_0^2)^{1/2},$$

$$D_{1,2} \approx [\omega^2 - k^2c^2 \mp 2\omega_0(\omega - kv_{go})] ,$$

$\chi_e \simeq -\omega_p^2/\omega^2$, $\chi_i \simeq -\omega_{pi}^2/\omega^2$, Eq. (7.62) can be rewritten as

$$\frac{\omega_p^2-\omega^2}{\omega_{pi}^2-\omega^2} = -\frac{\omega_p^2}{\omega^2}\frac{k^2|v_0|^2}{2}\left[\frac{\omega^2-k^2c^2}{(\omega^2-k^2c^2)^2-4\omega_0^2(\omega-kv_{go})^2}\right] . \tag{7.63}$$

It gives an unstable root for $\omega^2 < \omega_{pi}^2$; $\omega = kv_{go} + i\gamma$,

$$\gamma^2 = \frac{\omega_{pi}^2}{8\omega_0^2}\frac{\omega_p^2}{\omega_0^2}\frac{|v_0|^2}{c^2}\omega^2 , \tag{7.64}$$

where we have assumed $2\omega_0(\omega - kv_{go}) > \omega^2 - k^2c^2$ i.e., $\gamma > \omega_p^2\omega^2/2\omega_0^3$ requiring

$$\frac{|v_0|}{c} > \frac{\omega_p}{\omega_0}\frac{\omega}{\omega_{pi}}\sqrt{2} . \tag{7.65}$$

If one takes $\omega \leq \omega_{pi}\omega_0|v_0|/c\omega_p 2\sqrt{2}$ in compliance with Eq. (7.65) the growth rate can be written as

$$\gamma \leq \frac{\omega_{pi}^2}{\omega_0}\frac{|v_0|^2}{8c^2} .$$

which is rather small.

7.10. Filamentation Instability

In this case the low frequency density perturbation has **k** vector perpendicular to $\mathbf{k}_0$ so that the different beamlets of the pump wave front experience different indices of refraction. The beamlets converge towards refractive index maxima, i.e., into density depressions, enhancing the density perturbation. For $\mathbf{k} \perp \mathbf{k}_0$ we have $k_1^2 = k_2^2 = k^2 + k_0^2$ and $D_1 = D_2 = \omega^2 - k^2c^2/\omega_0^2$. Taking $kv_{th} > \omega > kv_{thi}$, $\chi_e \simeq \omega_{pi}^2/k^2c_s^2$, $\chi_i = -\omega_{pi}^2/\omega^2$ so that the plasma could quickly follow the variations in ponderomotive potential ϕ_p caused by the high frequency wave (the pump and the sidebands), the dispersion relation (7.62) can be cast as

$$\omega^2 - k^2c_s^2 = -\frac{|v_0|^2}{2c^2}(\omega_{pi}^2 - \omega^2) , \tag{7.66}$$

which gives a purely growing mode ($\omega - i\gamma$), when $|v_0|^2/c^2 > 2k^2c_s^2/\omega_{pi}^2$, with the growth rate

$$\gamma \simeq \frac{|v_0|}{c\sqrt{2}}\omega_{\rm pi} \,. \tag{7.67}$$

It is useful to obtain the spatial growth rate of the zero frequency mode. For $\omega = 0$, $\mathbf{k} = \mathbf{k}_\perp + k_z\hat{z}$ (where $\mathbf{k}_\perp$ is perpendicular to $\hat{z}$), Eq. (7.62) can be rewritten as

$$1 = \frac{\omega_{\rm pi}^2}{k^2c_{\rm s}^2}\frac{1}{1+T_{\rm i}/zT_{\rm e}}\frac{k^2|v_0|^2}{2}\frac{k^2c^2}{k^4c^4 - 4\omega_0^2k_z^2c^2}\,, \tag{7.68}$$

which gives the spatial growth rate $G = ik_z$,

$$G = \frac{k_\perp}{2}\left[\frac{\omega_{\rm p}^2}{\omega_0^2}\frac{|v_0|^2}{(1+T_{\rm i}/zT_{\rm e})v_{\rm th}^2} - \frac{k_\perp^2c^2}{\omega_0^2}\right]^{1/2} \tag{7.69}$$

where we have assumed $k_\perp^2 < k_z^2$. The first term inside the square root is due to nonlinear refraction whereas the second one or is due to diffraction. The diffraction gets stronger when filament size ($\sim k_\perp^{-1}$) is smaller, overpowering nonlinear refraction. The growth rate maximizes at

$$\begin{aligned} k_\perp = k_{\perp\rm opt} &= \frac{\omega_{\rm p}}{\sqrt{2}c}\frac{|v_0|}{v_{\rm th}}\frac{1}{(1+T_{\rm i}/zT_{\rm e})}\,, \\ G_{\max} &= \frac{\omega_0}{4c}\frac{|v_0|^2}{v_{\rm th}^2(1+T_{\rm i}/zT_{\rm e})}\frac{n_0^0}{c_{\rm cr}} \end{aligned} \tag{7.70}$$

This is the same expression as obtained in Chapter 6 for the case of ponderomotive nonlinearity.

7.11. Decay Instability

Near the critical layer an electromagnetic pump wave can decay into an ion acoustic wave and a Langmuir wave sideband. It is also probable that the incident p-polarized light may undergo linear mode conversion producing a large amplitude Langmuir wave which then decays into a pair of non acoustic and Langmuir waves.

A. *Electromagnetic pump*

Consider the decay of an electromagnetic pump $\mathbf{E}_0 = \hat{x}E_0\exp[-i(\omega_0 t - k_0 z)]$ into an ion acoustic wave $\phi = \phi\exp[-i(\omega t - \mathbf{k}\cdot\mathbf{x})]$ and a Langmuir wave $\phi_1 = \phi_1\exp[-i(\omega_1 t - \mathbf{k}_1\cdot\mathbf{x})]$, where $\omega_1 = \omega - \omega_0$, $\mathbf{k}_1 = \mathbf{k} - \mathbf{k}_0$. Since $\omega \ll \omega_{\rm pi}$, $|\omega_1|$ is nearly equal to ω_0. Further, $|\omega_1| \gg k_1 v_{\rm th}$ to ensure weak damping of the Langmuir wave, i.e., $|\omega_1| - \omega_{\rm p} \ll \omega_1$ hence $\omega_0 - \omega_{\rm p} \ll \omega_0$ and the parametric instability occurs near the critical layer. Writing $\omega_0 - \omega_p \simeq k_0^2c^2/2\omega_p$, $|\omega_1| - \omega_p \simeq 3k_1^2v_{th}^2/4\omega_p$, $\omega_3 \simeq kc_s$ in the phase matching condition $\omega = \omega_1 + \omega_0$ we obtain, $kc_s = \omega_0 - \omega_p - 3k_1^2v_{th}^2/4\omega_p$. Solving for k this equation gives

$$|k|c_s = \left[k_0 c_s \cos\Theta - \frac{1}{3}\left(\frac{m}{m_i}\right)^{1/2}\omega_{pi} + \sqrt{\left(k_0 c_s \cos\Theta - \frac{1}{3}\left(\frac{m}{m_i}\right)^{1/2}\omega_{pi}\right)^2 + \frac{1}{3}k_0^2c^2\frac{m}{m_i}}\right] . \tag{7.71}$$

As $k_0 \to 0$, $k \to 0$. For $k_0 > \omega_{pi}/3c$, $k \simeq \sqrt{2/3}k_0c/v_{th}$ i.e., the decay instability produces short wavelengths.

Now we proceed to calculate the nonlinear coupling. The high-frequency fields produce oscillatory velocity of electrons

$$\mathbf{v}_0 = \frac{e\mathbf{E}_0}{mi\omega_0}, \qquad \mathbf{v}_1 = -\frac{e\mathbf{k}_1\phi_1}{m\omega_1} \tag{7.72}$$

and exert a ponderomotive force on them $\mathbf{F}_p = e\nabla\phi_p$,

$$\phi_p = -\frac{m}{2e}\mathbf{v}_0 \cdot \mathbf{v}_1 = \frac{\mathbf{k}\cdot\mathbf{v}_0}{2\omega_1}\phi_1 . \tag{7.73}$$

In response to ponderomotive and self consistent potentials the low-frequency electron density perturbation can be written as

$$n = \frac{k^2}{4\pi e}\chi_e(\phi + \phi_p) \tag{7.74}$$

whereas the ion density perturbation is $n_i = -k^2\chi_i\phi/4\pi e$. Using n and n_i in the Poisson's equation,

$$\varepsilon\phi = -\chi_e\phi_p = -\frac{\mathbf{k}\cdot\mathbf{v}_0}{2\omega_1}\chi_e\phi_1 . \tag{7.75}$$

The low-frequency density perturbation n beats with $\mathbf{v}_0$ to produce a nonlinear density perturbation n_1^{NL} at the sideband. From the equation of continuity

$$-i\omega_1 n_1^{NL} + i k_1 \cdot \left(\frac{1}{2}n\mathbf{v}_0^*\right) = 0 \tag{7.76}$$

hence, the total density perturbation for the Langmuir sideband is

$$n_1 = \frac{k_1^2}{4\pi e}\chi_{1e}\phi_1 + \frac{\mathbf{k}\cdot\mathbf{v}_0^*}{2\omega_1}n , \tag{7.77}$$

where χ_{1e} is the electron susceptibility at $\omega_1, \mathbf{k}_1$ containing kinetic effects. Employing n_1 in the Poisson equation we get

$$\varepsilon_1\phi_1 = -\frac{4\pi e}{k_1^2}\frac{\mathbf{k}\cdot\mathbf{v}_0^*}{2\omega_1}n \simeq -\frac{k^2}{k_1^2}\frac{(1+\chi_i)\chi_e}{\varepsilon}\frac{|\mathbf{k}\cdot\mathbf{v}_0^*|^2}{4\omega_0^2}\phi_1 , \tag{7.78}$$

hence, the nonlinear dispersion relation:

$$\varepsilon\varepsilon_1 \simeq \frac{\omega_{\rm pi}^4}{\omega^2k^2c_s^2}\frac{|\mathbf{k}\cdot\mathbf{v}_0^*|^2}{4\omega_0^2} . \tag{7.79}$$

Expressing $\omega = \omega_r + i\gamma$, $\omega_1 = \omega_{1r} + i\gamma$ where $\omega_r, \omega_{1r} = (\omega_r - \omega_0)$ are the simultaneous zeros of $\varepsilon_r(\omega)$ and $\varepsilon_{1r}(\omega_1)$, the real parts of ε and ε_1, we write

$$\begin{aligned} \varepsilon &\simeq i\frac{\partial\varepsilon_r}{\partial\omega}(\gamma+\gamma_{\rm L}) , \\ \varepsilon_1 &\simeq i\frac{\partial\varepsilon_{1r}}{\partial\omega_1}(\gamma+\gamma_{\rm L1}) , \end{aligned} \tag{7.80}$$

where $\gamma_{\rm L} = \varepsilon_i/(\partial\varepsilon_r/\partial\omega)$, $\gamma_{\rm L1} = \varepsilon_{1i}/(\partial\varepsilon_{1r}/\partial\omega_1)$, are the linear damping rate of the decay waves and we have suppressed subscript r from ω, ω_1. Using Eqs. (7.80) Eq. (7.79) yields

$$(\gamma+\gamma_{\rm L})(\gamma+\gamma_{\rm L1}) = \gamma_{\rm M}^2 \tag{7.81}$$

where

$$\gamma_{\rm M}^2 = -\frac{\omega_{\rm pi}^4}{\omega^2k^2c_s^2}\frac{1}{\frac{\partial\varepsilon_r}{\partial\omega}\frac{\partial\varepsilon_{1r}}{\partial\omega_1}}\frac{|\mathbf{k}\cdot\mathbf{v}_0|^2}{4\omega_0^2} \approx \frac{\omega\omega_{\rm p}}{8}\frac{|\mathbf{k}\cdot\mathbf{v}_0|^2}{k^2v_{\rm th}^2} . \tag{7.82}$$

The threshold for the instability is given by $\gamma_{\rm M}^2 = \gamma_{\rm L}\gamma_{\rm L1}$. Much above the threshold the growth rate is $\gamma = \gamma_{\rm M}$. It maximizes for $\mathbf{k}||\mathbf{v}_0$.

In cases where the low frequency mode is strongly Landau damped on ions the decay process goes over to nonlinear Landau damping, giving a growth rate (cf. Eq. (7.78))

$$\gamma = -\gamma_{\rm L1} + \frac{\omega_{\rm p}}{2}\frac{|\mathbf{k}\cdot\mathbf{v}_0|^2}{k^2v_{\rm th}^2}\frac{\omega_{\rm pi}^2}{k^2c_s^2}\,{\rm Im}\left(-\frac{1}{\varepsilon}\right) . \tag{7.83}$$

B. *Electrostatic pump*

A large amplitude Langmuir wave can resonantly decay into the same pair of decay waves, mentioned in case A. However, since $|\omega_1| < \omega_0$, $|k_1| < k_0$, i.e., the decay produces longer-wavelength modes. Denoting the angle between $\mathbf{k}_0$ and $\mathbf{k}_1$ as Θ_1, the frequency of the low frequency mode can be written as $\omega \simeq k_0c_s(1 + \cos\Theta_1)$ which maximizes for back scatter to $\omega = 2k_0c_s$. The growth rate for this channel of decay is given by Eq. (7.82). For the case of nonlinear Landau damping γ is given by Eq. (7.83).

7.12. Oscillating Two-Stream Instability

Oscillating two-stream instability (OTSI) is a process by which a long wavelength pump wave excites short wavelength electrostatic waves and a purely growing density perturbation. The dynamics of the process can be understood as follows. Consider a dipole pump $\mathbf{E}_0 = \hat{x}E_0\cos\omega_0 t$, producing an oscillatory electron velocity $\mathbf{v}_0 = -\hat{x}(eE_0/m\omega_0)\sin\omega_0 t$. Let there also exist a small amplitude standing Langmuir wave $\mathbf{E}_1 = \hat{x}E_1\cos\omega_0 t\cos kx$. In zones $(2n-1/2)\pi < kx < (2n+1/2)\pi$, (where $n = 0, 1, 2, \ldots$) $\mathbf{E}_0$ and $\mathbf{E}_1$ are parallel to each other at all times. At other values of x they are antiparallel. The $\mathbf{E}_0$ and $\mathbf{E}_1$ fields exert a static ponderomotive force $\mathbf{F}_p = \hat{x}ed\phi_p/dx$, with $\phi_p = (eE_0E_1/2m\omega_0^2)\cos kx$, on the electrons, pushing plasma away from the zones where $\mathbf{E}_0$ and $\mathbf{E}_1$ are parallel, causing local density depression $\delta n \simeq (e\phi_p/T_e)n_0^0$ where $n_0^0, -e, m$ and T_e are unperturbed density, charge, mass and temperature of the electrons. δn in conjunction with velocity $\mathbf{v}_0$ produces a nonlinear current $\mathbf{J}_1^{\mathrm{NL}} = -\delta ne\mathbf{v}_0/2$ and a density perturbation $n_1^{\mathrm{NL}} = -(n_0^0ekv_0^2E_1/4T_e\omega_0^2)\cos\omega_0 t\sin kx$ where $v_0 = eE_0/m\omega_0$. The space charge perturbation n_1^{NL} produces an electric field $\mathbf{E}_1' = -\hat{x}\phi_1'/\partial x$ with $\phi_1' = -4\pi en_1/k^2\varepsilon_1$, i.e., $\mathbf{E}_1' = -\hat{x}(\pi n_0^0e^2V_0^2/\omega_0^2T_e\varepsilon_1)$ $E_1\cos\omega_0 t\cos kx$, where $\varepsilon_1 = 1 - \left(\omega_p^2 + 3k^2v_{\mathrm{th}}^2/2\right)/\omega_0^2$, and ω_p and v_{th} are electron plasma frequency and thermal speed. For $\mathbf{E}_1'$ to be in phase with $\mathbf{E}_1$, so that E_1 may grow with time, one must have $\varepsilon_1 < 0$, i.e., $\omega_0 < \omega_p\left(1 + 3k^2v_{\mathrm{th}}^2/2\omega_p^2\right)^{1/2}$. Further, for $\mathbf{E}_1'$ to have large magnitude ε_1 should be as small as possible, i.e., the frequency mismatch $\Delta = \omega_0 - \omega_p\left(1 + 3k^2v_{\mathrm{th}}^2/2\omega_p^2\right)^{1/2}$ should be of the order of the growth rate.

Consider an unmagnetized plasma with a coherent dipole pump

$$\mathbf{E}_0 = \hat{z}A_0e^{-i\omega_0 t} \tag{7.84}$$

The pump gives rise to oscillatory electron velocity

$$\mathbf{v}_0 = \frac{e\mathbf{E}_0}{mi\omega_0}\,. \tag{7.85}$$

We perturb this equilibrium by a low-frequency perturbation $(\omega, \mathbf{k})$ and two Langmuir wave sidebands $(\omega_{1,2}, \mathbf{k})$ where $\omega_{1,2} = \omega \pm \omega_0$. The electrostatic potentials are

$$\phi = \Phi e^{-i(\omega t - \mathbf{k}\cdot\mathbf{x})} ,$$
$$\phi_{1,2} = \Phi_{1,2} e^{-i(\omega_{1,2} t - \mathbf{k}\cdot\mathbf{x})} . \tag{7.86}$$

$\phi_{1,2}$, to the zeroth order, produces oscillatory velocities

$$\mathbf{v}_{1,2} = -\frac{e\mathbf{k}}{m\omega_{1,2}}\phi_{1,2} . \tag{7.87}$$

To the next order the sidebands and the pump exert a low-frequency ponderomotive force on the electrons $\mathbf{F}_p \equiv e\nabla\phi_p = -(m/2)\nabla(\mathbf{v}_0 \cdot \mathbf{v}_1 + \mathbf{v}_0^* \cdot \mathbf{v}_2)$, i.e.,

$$\phi_p = -\frac{ie\mathbf{k}}{2m\omega_0} \cdot \left(\frac{\mathbf{E}_0\phi_1}{\omega_1} - \frac{\mathbf{E}_0^*\phi_2}{\omega_2} \right) . \tag{7.88}$$

The ponderomotive and self-consistent potentials ϕ_p and ϕ produce an electron density perturbation

$$n = \frac{k^2}{4\pi e}\chi_e(\phi + \phi_p) , \tag{7.89}$$

where χ_e is the electron susceptibility at $\omega, \mathbf{k}$. For $\omega \ll kv_{\text{th}}$ (v_{th} being the electron thermal speed) $\chi_e = 2\omega_{\text{p}}^2/k^2v_{\text{th}}^2$. The ponderomotive force on the ions is down by the electron to ion mass ratio, hence, the ion density perturbation can be written as,

$$n_i = \frac{k^2}{4\pi e}\chi_i\phi , \tag{7.90}$$

where e is the ion charge, χ_i ($\simeq 2\omega_{\text{pi}}^2/k^2v_{\text{th}}^2$, for $\omega \ll kv_{\text{thi}}$) is the ion susceptibility and ω_{pi} and v_{thi} are the ion frequency and ion thermal speed. Using n and n_i in the Poisson equation $\nabla^2\phi = 4\pi e(n - n_i)$ we obtain

$$\varepsilon\phi = -\chi_e\phi_p , \tag{7.91}$$

where $\varepsilon \equiv 1 + \chi_e + \chi_i \simeq (\omega_{\text{pi}}^2/k^2c_s^2)(1 + T_{\text{e}}/T_{\text{i}})$, c_s is the sound speed and T_{e} and T_{i} are the electron and ion temperatures. One may rewrite this, using Eqs. (7.89)–(7.91)

$$n \simeq n_i = \frac{k^2}{4\pi e}\frac{\chi_i\chi_e}{\varepsilon}\phi_p . \tag{7.92}$$

The density perturbation n in conjunction with the oscillatory velocity $\mathbf{v}_0$ produces nonlinear density perturbations at $(\omega_{1,2}, \mathbf{k})$

$$n_1^{\rm NL} = \frac{\mathbf{k}\cdot\mathbf{v}_0^*}{2\omega_1}n = -\frac{\chi_i\chi_e}{\varepsilon}\frac{k^2\mathbf{k}\cdot\mathbf{E}_0^*\phi_p}{8\pi i m\omega_0\omega_1},$$
$$n_2^{\rm NL} = \frac{\mathbf{k}\cdot\mathbf{v}_0}{2\omega_2}n = -\frac{\chi_i\chi_e}{\varepsilon}\frac{k^2\mathbf{k}\cdot\mathbf{E}_0\phi_p}{8\pi i m\omega_0\omega_2}, \tag{7.93}$$

The linear density perturbations at the sidebands are

$$n_{1,2}^{\rm L} = \frac{k^2}{4\pi e}\chi_{e1,2}\phi_{1,2}\ , \tag{7.94}$$

where $\chi_{e1,2}$ are the electron susceptibilities at $(\omega_{1,2},\mathbf{k})$. Using Eqs. (7.93) and (7.94) in the Poisson equation we obtain $\varepsilon_j\phi_j = -4\pi e n_j^{\rm NL}/k^2$, $j=1,2$ or

$$\varepsilon_1\phi_1 = -\frac{e^2\mathbf{k}\cdot\mathbf{E}_0^*}{4m^2\omega_0^4}\frac{\chi_i\chi_e}{\varepsilon}(\mathbf{k}\cdot\mathbf{E}_0\phi_1 - \mathbf{k}\cdot\mathbf{E}_0^*\phi_2)\ ,$$
$$\varepsilon_2\phi_2 = \frac{e^2\mathbf{k}\cdot\mathbf{E}_0}{4m^2\omega_0^4}\frac{\chi_i\chi_e}{\varepsilon}(\mathbf{k}\cdot\mathbf{E}_0\phi_1 - \mathbf{k}\cdot\mathbf{E}_0\phi_2)\ , \tag{7.95}$$

where $\epsilon_j = 1-\left(\omega_{\rm p}^2+\frac{3}{2}k^2v_{\rm th}^2\right)/\omega_j^2$. Defining $\Delta = \omega_0 - \left(\omega_{\rm p}^2+\frac{3}{2}k^2v_{\rm th}^2\right)^{1/2}$, $a_0 = eA_0\cdot k_z/m\omega_0^2$, $\Omega = (\omega_0/8)(\chi_i\chi_e/\varepsilon) = (\omega_0/8)(\omega_{\rm pi}^2/k^2c_{\rm s}^2)\cdot(T_{\rm e}/T_{\rm e}+T_{\rm i})$, $c_{\rm s}^2 = T_{\rm e}/m_{\rm i}$ (where $m_{\rm i}$ is the ion mass) Eqs. (7.95) can be rewritten as

$$(\omega-\Delta-\Omega|a_0|^2)\Phi_1 = \Omega a_0^{*2}\Phi_2\ ,$$
$$(\omega+\Delta+\Omega|a_0|^2)\Phi_2 = -\Omega a_0^2\Phi_1\ , \tag{7.96}$$

giving the nonlinear dispersion relation

$$\omega^2 = [\Delta+\Omega|a_0|^2]^2 - \Omega^2|a_0|^4\ . \tag{7.97}$$

The instability occurs when frequency mismatch $\Delta < 0$. The maximum growth $\gamma_{\rm max}$ occur when $\Delta = -\Omega|a_0|^2$,

$$\gamma_{\rm max} = \Omega|a_0|^2 = \frac{|v_0|^2}{4v_{\rm th}^2}\frac{\omega_{\rm p}}{1+T_{\rm i}/T_{\rm e}}\ . \tag{7.98}$$

The optimum wave number is given by

$$k^2v_{\rm th}^2/\omega_{\rm p}^2 = \frac{4}{3}\left(\frac{\omega_0-\omega_{\rm p}}{\omega_{\rm p}}+\frac{|v_0|^2}{4v_{\rm th}^2}\right)\ .$$

If one introduces linear damping rates Γ_1, Γ_2 of the sideband waves in the parametric process, due to collisional or Landau damping, Eqs. (7.96) take the form

$$(\omega - \Delta - \Omega|a_0|^2 + i\Gamma_1)\phi_1 = \Omega a_0^{*2}\phi_2$$
$$(\omega + \Delta + \Omega|a_0|^2 + i\Gamma_2)\phi_2 = -\Omega a_0^2\phi_1 \tag{7.99}$$

giving (for $\Gamma_1 \simeq \Gamma_2$) a growth rate $\gamma \equiv -i\omega$,

$$\gamma = [\Omega^2|a_0|^4 - (\Delta + \Omega|a_0|^2)^2]^{1/2} - \Gamma_1 \,,$$
$$\gamma_{\max} = \Omega|a_0|^2 - \Gamma_1 \,. \tag{7.100}$$

The threshold for the instability is $|a_0|^2 \geq \Gamma_1/\Omega$. In a collisional plasma $\Gamma_1 \simeq \nu_{\rm ei}/2$. When $\nu_{\rm ei}$ is the electron-ion collision frequency.

7.13. Two-Plasmon Decay

Near quarter critical density ($n_0^0 \simeq n_{\rm cr}/4$) the pump electromagnetic wave excites a pair of Langmuir waves: $\phi = \phi\exp[-i(\omega t - \mathbf{k}\cdot\mathbf{x})]$, $\phi_1 = \phi_1\exp[-i(\omega_1 t - \mathbf{k}_1\cdot\mathbf{x})]$ where $\omega_1 = \omega - \omega_0$, $\mathbf{k}_1 = \mathbf{k} - \mathbf{k}_0$. Since all the three interacting waves are high-frequency modes, the motion of ions can be neglected. For the electron response one must solve the Vlasov equation since kinetic effects are important on both the decay waves. However, it would turn out that in the limit $\omega^2, \omega_1^2 \gg (k^2, k_1^2)v_{\rm th}^2$ the kinetic effects are not important in the nonlinear coupling term though very important in the dispersion and damping of the decay waves. Hence, we follow a cold fluid theory to evaluate the coupling coefficient. The linear response of electrons to ϕ is

$$\mathbf{v} = -\frac{e\mathbf{k}\phi}{m\omega} \,,$$
$$n = -\frac{n_0^0 e k^2\phi}{m\omega^2} \,. \tag{7.101}$$

The response $(\mathbf{v}_1, n_1)$ to ϕ_1 can be similarly written. Retaining the nonlinear terms in the equations of motion and continuity we obtain

$$\mathbf{v}^{\rm NL} = \frac{m}{2i\omega}[\mathbf{v}_0\cdot\nabla\mathbf{v}_1 + \mathbf{v}_1\cdot\nabla\mathbf{v}_0 + e\mathbf{v}_1\times\mathbf{B}_0/mc] = -\frac{e\mathbf{k}_1\cdot\mathbf{v}_0\phi_1}{2m\omega\omega_1}\mathbf{k} \,, \tag{7.102}$$

$$n^{\rm NL} = \frac{\mathbf{k}}{\omega}\cdot\left(n_0^0\mathbf{v}^{\rm NL} + \frac{1}{2}n_1\mathbf{v}_0\right) = -\frac{n_0^0 e\mathbf{k}\cdot\mathbf{v}_0}{2m\omega\omega_1}\left(\frac{k^2}{\omega} + \frac{k_1^2}{\omega_1}\right)\phi_1 \,. \tag{7.103}$$

The nonlinear density perturbation $n_1^{\rm NL}$ at ω_1, $\mathbf{k}_1$ can be obtained from $n^{\rm NL}$ by replacing $\mathbf{v}_0$, ϕ_1 by $\mathbf{v}_0^*$, ϕ. Using $n^{\rm NL}$ in the Poisson equation $\nabla^2\phi = 4\pi e(n^{\rm L} + n^{\rm NL})$ and employing $n^{\rm L} = k^2\chi_e\phi/4\pi e$ we obtain

$$\varepsilon\phi = \frac{\omega_p^2 \mathbf{k}\cdot\mathbf{v}_0}{2k^2\omega\omega_1}\left(\frac{k^2}{\omega}+\frac{k_1^2}{\omega_1}\right)\phi_1 \,. \tag{7.104}$$

Similarly one may write an equation for ϕ_1. Solving the coupled set we get

$$\varepsilon\varepsilon_1 \simeq \frac{|\mathbf{k}\cdot\mathbf{v}_0|^2}{4k^2k_1^2}\left(\frac{k^2}{\omega}+\frac{k_1^2}{\omega_1}\right)^2 \tag{7.105}$$

where $\varepsilon, \varepsilon_1$ are the dielectric functions at $\omega, \mathbf{k}$ and $\omega_1, \mathbf{k}_1$. Expanding $\varepsilon, \varepsilon_1$ around the resonant frequencies, i.e., taking $\omega = \omega + i\gamma$, $\varepsilon \simeq i(\gamma+\Gamma)\partial\varepsilon/\partial\omega$, $\varepsilon_1 \simeq i(\gamma+\Gamma_1)\partial\varepsilon_1/\partial\omega_1$ one obtains

$$(\gamma+\Gamma)(\gamma+\Gamma_1) = \gamma_M^2 \equiv \frac{|\mathbf{k}\cdot\mathbf{v}_0|^2}{16k^2k_1^2}[\mathbf{k}_0\cdot(\mathbf{k}_0+2\mathbf{k})]^2 \,. \tag{7.106}$$

For $\mathbf{k}_0 \| \hat{z}, \mathbf{E}_0 \| \hat{y}$ (no loss of generality) γ_M maximizes to $k_0|v_0|/2$ when $k_z = k_y = k_0/2$, $k_x = 0$. This growth rate is the same as that of stimulated Raman scattering at the quarter critical layer (cf. Eq. (7.53)). It must be noted here that the growth rate for two-plasmon decay vanishes for a dipole pump $k_0 = 0$. The decay occurs at densities slightly below the quarter critical. As one goes to lower densities the Landau damping of the decay waves becomes stronger, suppressing the instability for $kv_{\text{the}}/\omega_p \geq 0.3$ i.e., $(\omega_0/2 - \omega_p)/\omega_p \geq 0.1$.

7.14. Parametric Instability of a Random Pump

So far we have assumed the pump wave to be monochromatic. One may ask: what would happen when the pump has a frequency width $\Delta\omega_0$. As long as $\Delta\omega_0 < \gamma$, the growth rate, it really should not matter as all the frequency components of the pump wave can feed energy to the same pair of daughter waves. For $\Delta\omega_0 > \gamma$, only a narrow spectrum of the pump wave is in resonance with a pair of daughter waves, reducing the effective intensity of the pump participating in the parametric process to $I_{\text{eff}} = I_0\gamma/\Delta\omega_0$ (Fig. 7.8) since the growth rate of a parametric process goes as $I_{\text{eff}}^{1/2}$

$$\gamma = \gamma_M\left(\frac{I_{\text{eff}}}{I_0}\right)^{1/2}$$

or

$$\gamma = \frac{\gamma_M^2}{\Delta\omega_0} \tag{7.107}$$

where γ_M is the growth rate when the pump is monochromatic. Now we proceed to develop a simple treatment for a resonant three-wave parametric

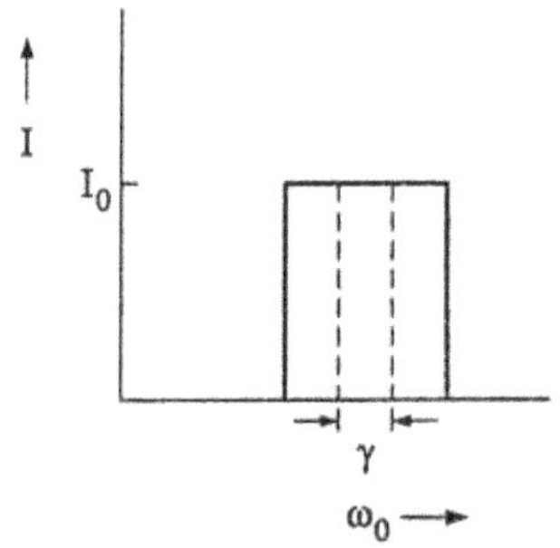

Fig. 7.8. Frequency spectrum of a broadband pump.

process with random pump. Neglecting damping the mode coupled equations for the temporal evolution of daughter waves of amplitudes a, a_1 can be written as

$$\frac{da}{dt} = \overline{\gamma}_M a_1 \tag{7.108}$$

$$\frac{da_1}{dt} = \overline{\gamma}_M^* a \ . \tag{7.109}$$

We take $\overline{\gamma}_M = |\overline{\gamma}_M| e^{i\phi}$ where ϕ is a random variable with a Gaussian distribution and its derivative $\dot{\phi} = f$ having a delta correlation function

$$\langle f(t) f(\tau) \rangle = \nu \delta(t - \tau) \ . \tag{7.110}$$

One may note that

$$\begin{aligned} \langle \dot{\phi} \rangle &= \langle f \rangle = 0 \ , \\ \frac{d}{dt}\langle \phi^2 \rangle &= 2\langle \phi \dot{\phi} \rangle = 2\left\langle f(t)\left[\int_0^t f(\tau) d\tau + \phi_0\right]\right\rangle \\ &= 2\int_0^t \langle f(t) f(\tau) \rangle d\tau = 2\nu \int_0^t \delta(t - \tau) d\tau = \nu \ , \end{aligned} \tag{7.111}$$

i.e., $\langle \phi \rangle$ = constant (can be taken to be zero) and $\langle \phi^2 \rangle = \nu t$. Thus ϕ indeed behaves in a random walker manner.

Equations (7.108) and (7.109) can be combined to give

$$\frac{d^2 a}{dt^2} - if\frac{da}{dt} = |\overline{\gamma}_M|^2 a \ . \tag{7.112}$$

This is a differential equation with a stochastic coefficient. To solve Eq. (7.112) we employ the "shot noise" model for f, i.e., writing f as a series of impulses

$$f = \sum_n \delta_n \delta(t - t_n) \tag{7.113}$$

where δ_n, the amplitude of each kick is small enough so that ϕ has a Gaussian distribution (by the central limit theorem). For $t \sim t_n$, the third term in Eq. (7.112) can be neglected. The solution is

$$\dot{a}(t) = \dot{a}(t_{n-}) \exp\left[i\delta_n \int_{t_{n-}}^{t} d\tau \delta(\tau - t_n) \right] \tag{7.114}$$

where $t_{n-} = t_n - \varepsilon$, $\varepsilon > 0$ and $\varepsilon \to 0$. We find that there is a slight shift in phase across each kick

$$\frac{\dot{a}(t_{n+})}{\dot{a}(t_{n-})} = e^{i\delta_n} . \tag{7.115}$$

To solve Eq. (7.112) we introduce a Green's function $G(t, \tau)$, satisfying equation

$$\frac{d^2 G}{dt^2} - |\overline{\gamma}_M|^2 G = \delta(t - \tau) \tag{7.116}$$

with $G(\tau, \tau) = 0$, $\dot{G}(t, \tau)|_{\tau-}^{\tau+} = 1$. $G(t, z)$ turns out to be

$$G(t, \tau) = \frac{1}{|\overline{\gamma}_M|} \sinh |\overline{\gamma}_M|(t - \tau) . \tag{7.117}$$

$a(t)$ can now be written as

$$a(t) = \frac{i}{\gamma_M} \int_{-\infty}^{t} \sinh[|\overline{\gamma}_M|(t - \tau)] f(\tau) \dot{a}(\tau) d\tau . \tag{7.118}$$

Eq. (7.118) can be rewritten as

$$a(t) = -\frac{i}{\gamma_M} \int_{\infty}^{0} \sinh |\overline{\gamma}_M| x f(t - x) \dot{a}(t - x) dx . \tag{7.119}$$

Differentiating Eq. (7.119) with respect to time and changing the variable again to $\tau = t - x$, we obtain

$$\dot{a}(t) - i \int_{-\infty}^{t} \cosh[|\overline{\gamma}_M|(t - t)] f(\tau) \dot{a}(\tau) d\tau . \tag{7.120}$$

The contribution to the integral comes from points $t = t_n$. Around these points $f(\tau)\dot{a}(\tau) \approx -i\ddot{a}$,

$$\begin{aligned}\dot{a}(t) &= \sum_n \int_{t_{n-}}^{t_{n+}} \cosh[|\overline{\gamma}_{\mathrm{M}}|(t-\tau)]\ddot{a}(\tau)d\tau \\ &\simeq \sum_{\substack{n \\ t_n<t}} \cosh[|\overline{\gamma}_{\mathrm{M}}|(t-t_n)][\dot{a}(t_{n+})-\dot{a}(t_{n-})] \\ &= \sum_{\substack{n \\ t_n<t}} \cosh[|\overline{\gamma}_{\mathrm{M}}|(t-t_n)](e^{i\delta_n}-1)\dot{a}(t_{n-}) \ . \end{aligned} \tag{7.121}$$

Since $\dot{a}(t_{n-})$ does not see the effect of δ_n (as δ_n occurs at a later time $t_n > t_{n-}$), the two are statistically independent, hence,

$$\langle \dot{a}(t)\rangle = \sum_{\substack{n \\ t_n<t}} \cosh[|\overline{\gamma}_{\mathrm{M}}|(t-t_n)]\langle e^{i\delta_n}-1\rangle\langle \dot{a}(t_{n-})\rangle \tag{7.122}$$

For a Gaussian distribution

$$\langle e^{i\delta_n}\rangle = 1 + \frac{\langle i\delta_n\rangle}{1!} + \frac{\langle (i\delta_n)^2\rangle}{2!} + \frac{\langle (i\delta_n)^3\rangle}{3!} + \ldots = e^{-\langle \delta_n^2\rangle/2} \equiv e^{-\overline{\nu}} \ . \tag{7.123}$$

Further, when t_n's are distributed uniformally at a separation $\Delta t, t_n = n\Delta t$. Using Eq. (7.113) one may write

$$\phi = \sum_n \delta_n, \qquad \langle \phi^2\rangle = \sum_n \langle \delta_n^2\rangle = 2n\overline{\nu} \ . \tag{7.124}$$

Earlier we have shown that $\langle \phi^2\rangle = \nu t$, hence $\overline{\nu} = (\nu/2)\Delta t$. Taking $e^{-\overline{\nu}} - 1 \simeq -(\nu/2)\Delta t$ the summation in Eq. (7.122) can be replaced by an integration,

$$\langle \dot{a}(t)\rangle = -\int_{-\infty}^{t} d\tau \cosh[|\overline{\gamma}_{\mathrm{M}}|(t-\tau)]\frac{\nu}{2}\langle \dot{a}\rangle \ . \tag{7.125}$$

Defining $Y \equiv \langle \dot{a}\rangle$ and differentiating Eq. (7.125) we obtain the non-stochastic equation,

$$\ddot{Y} = -\frac{\nu}{2}\dot{Y} + |\overline{\gamma}_{\mathrm{M}}|^2 Y \ . \tag{7.126}$$

Substituting $Y \simeq e^{\gamma t}$ we obtain

$$\gamma = \frac{1}{2}\left[-\frac{\nu}{2} + \sqrt{\frac{\nu^2}{4} + 4|\overline{\gamma}_{\mathrm{M}}|^2}\,\right] , \tag{7.127}$$

i.e.,

$$\gamma = |\overline{\gamma}_{\rm M}| \qquad \text{for } \nu < |\overline{\gamma}_{\rm M}|$$
$$= 2\frac{|\overline{\gamma}_{\rm M}|^2}{\nu} \qquad \text{for } \nu > |\overline{\gamma}_{\rm M}|;$$

ν is the frequency width or inverse coherence time of the pump. The frequency width of the pump thus suppresses the parametric instability. Recently there has been an upsurge of interest in using a broad band pump to suppress Raman, Brillouin and two-plasmon decay instabilities in a laser produced plasma. In the ISI (induced spatial incoherence) technique a laser beam is first divided into many independent beamlets which are given different time delays and then overlapped on the target. Experiments at Naval Research Laboratory[15] have been carried out irradiating disc targets with 2 ns pulses of 1.06 μ light. The Brillouin back scatter is severely reduced when the ISI echelons are used with either a medium or large bandwidth laser beam. Similar behavior is observed for the Raman scatter. However, it is unclear why there is such a huge reduction in SBS and SRS reflectivity while $\overline{\gamma}_{\rm M} \sim \nu$ in many of these experiments.

References

1. K. Nishikawa and C. S. Liu, in *Advances in Plasma Phys.*, eds. A. Simon and W. B. Thompson, **6**, 3 (Wiley, 1976).
2. C. S. Liu and P. K. Kaw, in *Advances in Plasma Phys.*, eds. A. Simon and W. B. Thompson, **6**, 83 (Wiley, 1976).
3. W. L. Kruer, *The Physics of Laser Plasma Interactions* (Addison-Wesley, 1987).
4. J. F. Drake, P. K. Kaw, Y. C. Lee, G. Schmidt, C. S. Liu, and M. N. Rosenbluth, *Phys. Fluids* **17**, 778 (1974).
5. B. I. Cohen and C. E. Max, *Phys. Fluids* **22**, 1115 (1979).
6. V. P. Silin, *Sov. Phys. JETP* **21**, 1127 (1965).
7. K. Nishikawa, *J. Phys. Soc. Japan* **24**, 1152 (1968).
8. D. W. Forslund, J. M. Kindel, and E. L. Lindman, *Phys. Fluids* **18**, 1002 (1975).
9. L. M. Gorbunov, *Soc. Phys. Usp.* **16**, 217 (1973).
10. C. S. Liu and V. K. Tripathi, *Phys. Fluids* **29**, 4188 (1986).
11. R. P. Drake, *et al.*, *Phys. Fluids* **31**, 1795 (1989); *ibid.* **31**, 3130 (1989).
12. W. L. Kruer, *UCRL report* 99302 (1988); also published in *Laser Plasma Interaction-4* ed. M. B. Hooper, (SUSSP, Edinburgh, 1989), p. 89.
13. H. A. Rose, D. F. DuBois, and B. Bezzerides, *Phys. Rev. Lett.* **58**, 2547 (1987).
14. C. J. Walsh, D. M. Villeneuve, and H. A. Baldis, *Phys. Rev. Lett.* **53**, 1445 (1984).
15. A. N. Mostovych *et al.*, *Phys. Rev. Lett.* **59**, 1193 (1987).

CHAPTER 8

A NONLINEAR SCHRÖDINGER EQUATION

We have seen that a large amplitude pump wave is unstable to modulational instability. The instability would have a large growth rate when the group velocity of the pump is comparable to or less than the speed of sound ($v_{go} < c_s$). One may wish to ask: to what level can the amplitude modulation build up? What is the amplitude envelope where the tendency for nonlinear build-up is balanced by dispersion broadening? These nonlinear effects lead to intensity phenomena of soliton formation and wave collapse. In this chapter we address these issues following the treatment of Nishikawa and Liu.[1]

8.1. Basic Equation

Consider the propagation of a Langmuir wave, with some kind of amplitude modulation, in a plasma

$$\phi = \phi_0(\mathbf{x}, t)e^{-i(\omega_0 t - k_0 z)} \tag{8.1}$$

where ϕ_0 is a slowly varying function of $\mathbf{x}$ and t and

$$\omega_0^2 = \omega_p^2 + \frac{3}{2}k_0^2 v_{th}^2 \,. \tag{8.2}$$

We restrict to values of ω_0 close to ω_p so that $\partial\omega_0/\partial k_0 < c_s$. The spatial nonuniformity in the amplitude of the wave exerts a ponderomotive force on the electrons $\mathbf{F}_p = -(m/2)\nabla(\mathbf{v}_0 \cdot \mathbf{v}_0^*) \equiv e\nabla\phi_p$ where $\mathbf{v}_0 = -e\mathbf{k}_0\phi_0/m\omega_0$,

$$\phi_p = -\frac{ek_0^2|\phi_0|^2}{2m\omega_0^2} \ . \tag{8.3}$$

Under this force the plasma undergoes ambipolar diffusion. In the quasi-steady state the pressure gradient force balances the ponderomotive force,

$$e\nabla\phi_p = (T_e + T_i)\nabla \ln n$$

giving

$$n = Ae^{e\phi_p/(T_e+T_i)} \simeq A\left[1 + \frac{e\phi_p}{T_e + T_i}\right] \ . \tag{8.4}$$

The density of the plasma averaged over space is, however, n_0^0,

$$n_0^0 = A\left[1 + \frac{e\langle\phi_p\rangle}{T_e + T_i}\right] \ , \tag{8.5}$$

where $\langle \ \rangle$ denotes the spatial average. Hence, using Eqs. (8.3), (8.4) and (8.5) we have

$$n = n_0^0 + \Delta n; \qquad \Delta n = -\frac{n_0^0}{2}\frac{|v_0|^2 - \langle|v_0^2|\rangle}{c_s^2} \ . \tag{8.6}$$

Incorporating the modified density (due to self action) in Eq. (8.2) and replacing $\omega_0 \to \omega_0 + i\partial/\partial t$, $\mathbf{k}_0 = \mathbf{k}_0 - i\partial/\partial\mathbf{x}$ we deduce the wave equation governing ϕ_0:

$$i\left(\frac{\partial\phi_0}{\partial t} + v_g\frac{\partial\phi_0}{\partial z}\right) + \frac{3}{4}\frac{v_{\rm th}^2}{\omega_0}\nabla^2\phi_0 - \frac{\omega_{\rm p}}{2}\frac{\Delta n}{n_0^0}\phi_0 = 0 \ , \tag{8.7}$$

where $\partial^2/\partial t^2 \sim v_g^2(\partial^2/\partial z^2)$ has been neglected as compared to $v_{\rm th}^2\partial^2/\partial z^2$ because $v_g \leq c_s \ll v_{\rm th}$. We introduce a new set of variables

$$t' = t, \qquad z' = z - v_g t, \qquad x' = x, \qquad y' = y \ ,$$

to refer to quantities in the moving frame, and eventually suppressing the prime in favor of brevity, Eq. (8.7) can be cast into the form of a nonlinear Schrödinger equation[1–4]

$$i\frac{\partial}{\partial t}\phi_0 + p\nabla^2\phi_0 + q(|\phi_0|^2 - \phi_{00}^2)\phi_0 = 0 \tag{8.8}$$

where

$$p = \frac{3v_{\rm th}^2}{4\omega_0}, \qquad q = \frac{\omega_{\rm p}}{4}\frac{e^2k_0^2}{m^2\omega_0^2c_s^2}, \qquad \phi_{00} = \langle|\phi_0|^2\rangle \ . \tag{8.9}$$

We begin by examining the stability of a plane uniform Langmuir wave to one-dimensional modulation perturbation

$$\phi_0 = \phi_{00} + \phi_1(z,t), \qquad \phi_1 = U + iV \tag{8.10}$$

Using Eq. (8.10) in Eq. (8.8) and separating real and imaginary parts we obtain

$$\begin{aligned} -\frac{\partial V}{\partial t} + p\frac{\partial^2 U}{\partial z^2} + 2q\phi_{00}^2 U = 0 \\ \frac{\partial U}{\partial t} + p\frac{\partial^2 V}{\partial z^2} = 0 \,. \end{aligned} \tag{8.11}$$

Taking $U, V \sim e^{-i(\Omega t - kz)}$, Eqs. (8.11) yield the nonlinear dispersion relation

$$\Omega^2 = p^2 k^4 - 2pq\phi_{00}^2 k^2 \tag{8.12}$$

An unstable solution ($\mathrm{Im}\,\Omega > 0$) occurs when $pq > 0$ and

$$0 < k^2 < 2q\phi_{00}^2/p \,. \tag{8.13}$$

Maximum growth occurs for $k = (q\phi_{00}^2/p)^{1/2}$, with a rate

$$\gamma_{\max} = q\phi_{00}^2 \,. \tag{8.14}$$

8.2. Stationary Solution

As the modulation grows, higher k's are produced by harmonic generation, enhancing the dispersion effect and tending to suppress the instability. A stationary state is realized as a balance of the nonlinear bunching effect and the dispersion effect. The nonlinear effect may produce a frequency shift, hence, we attempt a stationary solution,

$$\phi_0 = A(z)e^{-i\omega t} \,. \tag{8.15}$$

Using this in Eq. (8.8) we get

$$p\frac{d^2 A}{dz^2} + q\left[|A|^2 - \phi_{00}^2 + \frac{\omega}{q}\right] A = 0 \,. \tag{8.16}$$

Introducing an eikonal

$$A = A_0 e^{iS} \,, \tag{8.17}$$

Eq. (8.16) can be separated into real and imaginary parts

$$p\left[\frac{d^2A_0}{dz^2}-\left(\frac{ds}{dz}\right)^2A_0\right]+q\left[A_0^2-\phi_{00}^2+\frac{\omega}{q}\right]A_0=0\ , \tag{8.18}$$

$$\frac{ds}{dz}\frac{dA_0^2}{dz}+A_0^2\frac{d^2s}{dz^2}=0 \tag{8.19}$$

Eq. (8.19) yields

$$A_0^2\frac{ds}{dz}\equiv M=\text{const.} \tag{8.20}$$

Using Eq. (8.20) in and Eq. (8.19), multiplying the latter by $2(dA_0/dz)$ and integrating over z we obtain

$$\left(\frac{dA_0^2}{dz}\right)^2+4M^2+\frac{4A_0^4}{p}\left(\omega-q\phi_{00}^2+q\frac{A_0^2}{2}\right)-EA_0^2=0\ , \tag{8.21}$$

where E is a constant. Equation (8.21) has a general solution in terms of Jacobi's elliptic function

$$A_0^2=c_2+(c_1-c_2)cn^2\left[\left(\left|\frac{q}{p}\right|\frac{c_1-c_3}{6}\right)^{1/2}z,\chi\right]\ , \tag{8.22}$$

where c_1, c_2, c_3 ($c_1\geq c_2\geq c_3$) are the three solutions of the equation

$$2\frac{q}{p}c^3+\frac{4}{p}(\omega-q\phi_{00}^2)c^2-Ec+4M^2=0 \tag{8.23}$$

and

$$\chi=\left[\frac{c_1-c_2}{c_1-c_3}\right]^{1/2}\ .$$

A_0^2 is a periodic function of period $2K(\chi)$ where $K(\chi)$ is the complete elliptic integral of first kind of modulus χ. It is called a periodic wave train. In the special case where $\chi\to 1$ we get solitary wave solutions:

$pq\leq 0$:

$$A_0^2=\phi_{00}^2[1-a\ \text{sech}^2 bz]\ ;$$

$$a=-\frac{2pb^2}{q\phi_{00}^2},\qquad b=\left[-\frac{\omega}{p}-\frac{q}{2p}\phi_{00}^2\right]^{1/2}\ , \tag{8.24}$$

$pq\geq 0$:

$$A_0^2=-\frac{2\omega}{q}\text{sech}^2\left[\left(-\frac{\omega}{p}\right)^{1/2}z\right],\qquad M=0\ . \tag{8.25}$$

These solitary wave solutions are called envelope solitons.

8.3. Instability of an Envelope Soliton

Now we examine the stability of an envelope soliton against a perpendicular perturbation, i.e., against filamentation,

$$\phi_0 = [A(z) + a_1(x, z, t)]e^{-i\omega t} , \tag{8.26}$$

where $A = a_0 \operatorname{sech}\left[(q/2p)^{1/2} a_0 z\right]$, $a_0 = (-2\omega/q)^{1/2}$. It may be realized here that the soliton is localized spatially, hence $\phi_{00}^2 \approx 0$ (ϕ_{00}^2 representing the average value of $|\phi_0^2|$ over a length in which plasma can undergo ambipolar diffusion in one period $2\pi\omega^{-1}$) as long as $(q/2p)^{1/2} a_0 c_s > q(a_0^2/2)$. Using Eq. (8.26) in Eq. (8.8) one obtains on linearization

$$\omega a_1 + i\frac{\partial a_1}{\partial t} + p\frac{\partial^2 a_1}{\partial z^2} + p\frac{\partial^2 a_1}{\partial x^2} + qA^2(2a_1 + a_1^*) = 0 . \tag{8.27}$$

Expressing $a_1 = u + iv$, Eq. (8.27) can be separated into real and imaginary parts:

$$-\frac{\partial v}{\partial t} + \left[\omega + p\frac{\partial^2}{\partial z^2} + p\frac{\partial^2}{\partial x^2} + 3qA^2\right] u = 0 \tag{8.28}$$

$$\frac{\partial u}{\partial t} + \left[\omega + p\frac{\partial^2}{\partial z^2} + p\frac{\partial^2}{\partial x^2} + qA^2\right] v = 0 . \tag{8.29}$$

Multiplying Eq. (8.28) by $\partial A/\partial z$ and Eq. (8.29) by A and carrying out the z integration from $-\infty$ to ∞, we obtain

$$-\frac{\partial}{\partial t}\int dz\frac{\partial A}{\partial z}v + p\frac{\partial^2}{\partial x^2}\int dz\frac{\partial A}{\partial z}u = 0 , \tag{8.30}$$

$$\frac{\partial}{\partial t}\int dzAu + p\frac{\partial^2}{\partial x^2}\int dzAv = 0 , \tag{8.31}$$

where we have employed Eq. (8.16) with $\phi_{00}^2 = 0$ and carried out integration by parts of the third terms in Eq. (8.28) and (8.29).

If one views the perturbation to represent a modification in the amplitude of the soliton, i.e., $a_0 \to a_0 + \delta a_0$, then using the definition of A and ω, ϕ_0 may be written as

$$\phi_0 \simeq Ae^{-i\omega t} + \frac{\partial}{\partial a_0}\left(Ae^{iq\frac{a_0^2}{2}t}\right)\delta a_0$$

$$= \left\{A + \delta a_0 \frac{\partial}{\partial z}\left[z \sec h\left(\left(\frac{q}{2p}\right)^{1/2} a_0 z\right)\right]\right.$$

$$\left. + iqa_0^2 \sec h\left(\left(\frac{q}{2p}\right)^{1/2} a_0 z\right)\int^t \delta a_0 dt\right\} e^{-i\omega t} . \tag{8.32}$$

The real and imaginary parts of a_1 can now be written as

$$u = \delta a_0 \frac{\partial}{\partial z}\left[z \sec h\left(\left(\frac{q}{2p}\right)^{1/2} a_0 z\right)\right]$$

$$v = qa_0^2 \sec h\left(\left(\frac{q}{2p}\right)^{1/2} a_0 z\right)\int^t \delta a_0 dt . \tag{8.33}$$

Substituting these expressions for u and v in Eq. (8.31) and differentiating the result with respect to time, we obtain

$$\frac{\partial^2}{\partial t^2}(\delta a_0) + 2pqa_0^2 \frac{\partial^2}{\partial x^2}(\delta a_0) = 0 . \tag{8.34}$$

For $pq > 0$ the solution to this equation is unstable. A similar procedure can be followed to examine a perturbation in the speed of the soliton which turns out to be unstable when $pq < 0$. Thus the envelope soliton is always unstable either against the amplitude perturbation (granulation) or against the velocity perturbation (flutter).

8.4. Criterion for Collapse

We may deduce some general conclusions on the stability of a multidimensional nonlinear Schrödinger equation. Ignoring ϕ_{00}^2 term, Eq. (8.8) can be written in the form of a Schrödinger equation

$$i\frac{\partial}{\partial t}\phi_0 = H\phi_0 \tag{8.35}$$

with the Hamiltonian

$$H = -p\nabla^2 - q|\phi_0|^2 . \tag{8.36}$$

The localized solution therefore satisfies two conservation laws:

$$\text{“Density” Conservation:} \quad \int |\phi_0|^2 d^3\mathbf{x} = I_1 = \text{const.} \tag{8.37}$$

$$\text{“Energy” Conservation:} \quad \int \left(p|\nabla\phi_0|^2 - \frac{1}{2}q|\phi_0|^4\right) d^3\mathbf{x}$$

$$= I_2 = \text{const.} \tag{8.38}$$

Now we investigate the time evolution of the spatial variance (or size of the soliton)

$$\langle|\mathbf{x}|^2\rangle = \frac{1}{I_1}\int d^3\mathbf{x}|\mathbf{x}|^2|\phi_0|^2 \,. \tag{8.39}$$

Introducing momentum operator $\mathbf{P} = -i\nabla$ and using the equations of motion

$$\frac{d\mathbf{x}}{dt} = 2p\mathbf{P}, \qquad \frac{d\mathbf{P}}{dt} = q\nabla|\phi_0|^2 \tag{8.40}$$

we can write

$$\frac{d}{dt}\langle|\mathbf{x}|^2\rangle = \left\langle \mathbf{x}\cdot\frac{d\mathbf{x}}{dt} + \frac{d\mathbf{x}}{dt}\cdot\mathbf{x}\right\rangle = 2p\langle\mathbf{x}\cdot\mathbf{p} + \mathbf{p}\cdot\mathbf{x}\rangle \tag{8.41}$$

$$\begin{aligned}\frac{d^2\langle|\mathbf{x}|^2\rangle}{dt^2} &= \left\langle \frac{d\mathbf{x}}{dt}\cdot\mathbf{p} + \mathbf{x}\cdot\frac{d\mathbf{P}}{dt} + \mathbf{P}\cdot\frac{d\mathbf{x}}{dt} + \frac{d\mathbf{P}}{dt}\cdot\mathbf{x}\right\rangle \\ &= 4p\langle 2p|\mathbf{P}|^2 + q\mathbf{x}\cdot\nabla|\phi_0|^2\rangle \\ &= \frac{4P}{I_1}\int d^3\mathbf{x}\left\{2p|\nabla\phi_0|^2 + \frac{q}{2}\mathbf{x}\cdot\nabla|\phi_0|^4\right\} \,,\end{aligned} \tag{8.42}$$

where we have used $\langle s\rangle = (1/I_1)\int \phi_0^* s\phi_0 d^3\mathbf{x}$,

$$\int \phi_0^*\frac{\partial^2}{\partial x^2}\phi_0 dx = -\int \frac{\partial\phi_0}{\partial x}\frac{\partial\phi_0^*}{dx}dx = -\int\left|\frac{\partial\phi_0}{\partial x}\right|^2 dx \,.$$

Integrating the last term of Eq. (8.42) by parts

$$\frac{q}{2}\int \mathbf{x}\cdot\nabla|\phi_0|^4 d^3\mathbf{x} = -\frac{qd}{2}\int|\phi_0|^4 d^3\mathbf{x}$$

where $d = 3$ for a three-dimensional case (d stands for the dimension of the problem), Eq. (8.42) can be written as

$$\frac{d^2}{dt^2}\langle|\mathbf{x}|^2\rangle = \frac{4p}{I_1}\left[2I_2 + q\left(1 - \frac{d}{2}\right)\int d^3\mathbf{x}|\phi_0|^4\right] \,.$$

When $pI_2 > 0$ then the right hand side is always negative for $d \geq 2$. The solution then becomes

$$\langle |\mathbf{x}|^2 \rangle \leq c_1 + c_2 t + \frac{8p}{I_1} I_2 t^2$$

which collapses to zero at finite time.

The collapse of the plasma wave leads to caviton formation in the plasma and acceleration of charged particles. Dynamics of this collapse has been intensively studied theoretically and computationally with particle simulations, for instance, in Refs. 5–8.

References

1. K. Nishikawa and C. S. Liu, *Adv. in Plasma Physics*, eds. A. Simon and W. B. Thompson (Wiley, 1976) **6**, p. 3.
2. R. Z. Sagdeev and A. A. Galeev, *Nonlinear Plasma Theory* (Benjamin, New York, 1969).
3. V. E. Zakharov and A. B. Shabat, *Sov. Phys. JETP* **34**, 62 (1972); V. E. Zakharov, *Sov. Phys. JETP* **35**, 908 (1972).
4. H. H. Chen and C. S. Liu, *Phys. Rev. Lett.* **37**, 693 (1976).
5. A. A. Galeev, R. Z. Sagdeev, Yu S. Sigov, V. D. Shapiro, V. I. Shevchenko, *Sov. J. Plasma Phys.* **1**, 5 (1975).
6. V. E. Zakharov, A. N. Pushkarev, A. M. Rubenchik, R. Z. Sadeev and V. F. Shvets, *JETP Lett.* **47**, 287 (1988).
7. P. L. Newman, R. M. Winglee, P. A. Robinson, J. Glanz and M. N. Goldman, *Phys. Fluids* **B2**, 2600 (1991).
8. D. DuBois, H. A. Rose, D. Russell *Physica Scripta*, T30, 137–158 (1990).

CHAPTER 9

PARAMETRIC INSTABILITIES IN AN INHOMOGENEOUS PLASMA

The process of parametric interaction is strongly modified by the presence of plasma inhomogeneity.[1-8] The most important effects arise through the nonuniformity in plasma density, hence, we consider a plasma with primarily a density gradient, $\nabla n_0 \| \hat{x}$. For a pump wave launched from the low density side at an angle θ_0 to the density gradient, we may divide the plasma into four regions of interest:

(1) Underdense region, $n_0 \ll n_{0\mathrm{cr}}$,
(2) Quarter critical density layer, $n_0 \sim n_{0\mathrm{cr}}/4$,
(3) Turning point, $n_0 \sim n_{0\mathrm{cr}} \cos^2 \theta_0$,
(4) Critical layer, $n_0 \sim n_{0\mathrm{cr}}$.

In the underdense region stimulated Raman and Brillouin scattering are two prominent parametric processes with electromagnetic sidebands of frequency $\omega_1 \gg \omega_\mathrm{p}$. As long as the decay waves have a significant component of their wavevector along the density gradient (i.e., unless they propagate almost perpendicular to the density gradient) one may employ a WKB approximation to discuss the scattering processes. Since the **k** matching conditions ($k_{0x}(x) = k_x(x) - k_{1x}(x)$ for a plasma with $\nabla n_0 \| \hat{x}$) are satisfied only locally, the region of parametric interaction is localized. The propagation of energy by the decay waves out of this region sets a threshold for the growth of the instability. The instability turns out to be convective.

At large scattering angles, with respect to density gradient, the side-band electromagnetic wave has a turning point in the interaction region, allowing a sufficiently long time for nonlinear interaction. Consequently the instability could become absolute. However, with pump waves of narrow transverse extent convective losses sideways could be severe.

Near the quarter critical layer the frequency of Raman backscatter is close to local plasma frequency ($\omega_1 \sim \omega_p$), i.e., the back scattered electromagnetic wave has a turning point leading to larger convective amplification or absolute instability. In this region two-plasmon decay is an important parametric instability. The Langmuir waves have turning points and the process turns out to be an absolute instability.

In the region between the quarter critical and the turning point stimulated Brillouin scattering is the only three-wave parametric process. Near the turning point the pump wave acquires large amplitude and the growth rate of the instability is considerably enhanced. When the angle of incidence θ_0 is small, i.e., $\cos^2\theta_0 - 1 \ll 1$, one may excite decay instability and oscillating two stream instability. Further, if the pump wave is p-polarized, and the separation between the turning point and critical layer is small, large amplitude Langmuir waves are produced near the critical layer via linear mode conversion. These waves excite decay and oscillating two-stream instabilities with large growth rates.

9.1. WKB Solution

For parametric processes in the underdense region of an inhomogeneous plasma, one may replace $k_x \to k_x - i\partial/\partial x$ and $\omega \to \omega + i\partial/\partial t$. For Raman scattering, we may write the dielectric function for the plasma wave $\varepsilon(\omega, \mathbf{k})$ and $D_1(\omega_1, \mathbf{k}_1)$ for the scattered light in (7.46) as

$$\varepsilon(\omega, \mathbf{k}) = i\frac{\partial\varepsilon}{\partial\omega}\left(\frac{\partial}{\partial t} + v_{gx}\frac{\partial}{\partial x} + \Gamma\right)$$

$$D_1(\omega_1, \mathbf{k}_1) = i2\omega_1\left(\frac{\partial}{\partial t} + v_{g1x}\frac{\partial}{\partial t} + \Gamma_1\right)$$

to study the slow space-time variations of field amplitudes. Then Eqs. (7.43) and (7.46) take the form (for $\delta_1 = 0$)

$$\frac{\partial\tilde{\phi}}{\partial t} + v_{gx}\frac{\partial\tilde{\phi}}{\partial x} + \Gamma\tilde{\phi} = \gamma_M \frac{v_0}{|v_0|}\tilde{E}_1 e^{-iK'x^2/2} \tag{9.1}$$

$$\frac{\partial\tilde{E}_1}{\partial t} + v_{g1x}\frac{\partial\tilde{E}_1}{\partial x} + \Gamma_1\tilde{E}_1 = \gamma_M \frac{v_0^*}{|v_0|}\tilde{\phi} e^{iK'x^2/2} \tag{9.2}$$

where

$$\begin{aligned} \tilde{\phi} &= \frac{k}{4(\pi\hbar)^{1/2}}\left(\frac{\partial\varepsilon}{\partial\omega}\right)^{1/2}\phi \,, \\ \tilde{E}_1 &= \frac{1}{(8\pi\hbar)^{1/2}|\omega_1|^{1/2}}E_1 \,, \\ \gamma_M &= \frac{\chi_e k|v_0|}{2\sqrt{2}\left(|\omega_1|\frac{\partial\varepsilon}{\partial\omega}\right)^{1/2}} \,, \\ K' &= \left.\frac{dK}{dx}\right|_{x=0}, \quad \text{and } K = k_x(x) - k_0(x) - k_1(x) \,, \end{aligned} \tag{9.3}$$

$\tilde{\phi}$ and $\tilde{E}_1$ are the normalized amplitudes of the electrostatic and side band waves such that $|\tilde{\phi}|^2$ gives the plasmon density (action) and $|\mathbf{E}_1|^2$ gives the photon density, K is the wave number mismatch and we have chosen $x = 0$ as the point where phase matching is satisfied. The exponential factors on the RHS are originally $e^{\mp i\int K dx}$. On expanding K around $x = 0$ these simplify to give the said result. Usually the linear damping rate Γ_1 of the sideband is quite small and we ignore it. Expressing

$$\begin{aligned} \tilde{\phi} &= a e^{-iK'x^2/4} \\ \tilde{E}_1 &= a_1 e^{+iK'x^2/4} \,, \end{aligned} \tag{9.4}$$

Eqs. (9.1) and (9.2) can be written as

$$\left(\frac{\partial}{\partial t} + v_{gx}\frac{\partial}{\partial x} - iv_{gx}K'\frac{x}{2} + \Gamma\right)a = \overline{\gamma}_M a_1 \tag{9.5}$$

$$\left(\frac{\partial}{\partial t} + v_{g1x}\frac{\partial}{\partial x} + iv_{g1x}K'\frac{x}{2}\right)a_1 = \overline{\gamma}_M^* a \tag{9.6}$$

where $\overline{\gamma}_M = \gamma_M v_0/|v_0|$. For a steady state solution, we set $\partial/\partial t = 0$ in Eqs. (9.5)–(9.6). A simple case is when the damping of the low frequency mode is large, $\Gamma - iv_{gx}K'x/2 \gg v_{gx}\partial/\partial x$, and Eqs. (9.5), (9.6) become

$$\frac{da_1}{dx} = \left[\frac{\gamma_M^2}{(\Gamma - iv_{gx}K'x/2)v_{g1x}} - iK'\frac{x}{2}\right] a_1 \ . \tag{9.7}$$

For a sideband wave coming from $-\infty (v_{g1x} > 0)$ with $a_1 = a_{10}$, Eq. (9.7) can be integrated to obtain the amplitude after passing through the interaction region,

$$|a_1|_{x\to\infty} = a_{10}e^{\pi A} \tag{9.8}$$

where

$$A = \frac{\gamma_M^2}{|v_{gx}v_{g1x}K'|} \tag{9.9}$$

is the convective amplification factor. The linear damping of the low-frequency mode limits the height of the resonant interaction but broadens the width of the interaction region, resulting in an amplification independent of Γ. For parametric instability to be significantly amplified one must have $A \geq 1$, which determines the threshold pump power.

It can be shown that the temporal growth of the instability, in the WKB limit considered here, is not possible. Following Rosenbluth let us assume a and a_1 to vary as e^{pt} and write Eqs. (9.5) and (9.6) as

$$\frac{\partial a}{\partial x} + \left(\frac{p}{v_{gx}} - iK'\frac{x}{2}\right) a = \frac{\overline{\gamma}_M}{v_{gx}} a_1 \ , \tag{9.10}$$

$$\frac{\partial a_1}{\partial x} + \left(\frac{p}{v_{g1x}} - iK'\frac{x}{2}\right) a_1 = \frac{\overline{\gamma}_M}{v_{g1x}} a \ . \tag{9.11}$$

Here we have dropped Γ, which could be included trivially. Equations (9.10) and (9.11) combine to give an equation for a_1 which on expressing

$$a_1 = \psi(x)\exp\left[-p\frac{x}{2}\left(\frac{1}{v_{gx}} + \frac{1}{v_{g1x}}\right)\right] \tag{9.12}$$

leads to

$$\frac{d^2\psi}{dx^2} + \left[\left\{\frac{K'x}{2} + i\frac{p}{2}\left(\frac{1}{v_{gx}} - \frac{1}{v_{g1x}}\right)\right\}^2 - \frac{\gamma_M^2}{v_{gx}v_{g1x}} + i\frac{K'}{2}\right]\psi = 0 \ . \tag{9.13}$$

This is a parabolic cylinder equation. At large x, its solutions go as

$$\psi \sim \exp\left[\pm iK'\frac{x^2}{4} \mp \frac{px}{2}\left(\frac{1}{v_{gx}} - \frac{1}{v_{g1x}}\right)\right] \ , \tag{9.14}$$

hence $|a_1|$ goes as $e^{-px/v_{gx}}$ or $e^{-px/v_{g1x}}$. For Re(p) $\neq 0$ these solutions are badly behaved at $x = -\infty$ or $x = \infty$, depending on the signs of v_{gx} and v_{g1x}. Thus Re(p) must vanish for any physical solution, i.e., the temporal growth of the instability is not possible. One could solve Eq. (9.13) to calculate spatial amplification. It turns out to be the same as given by Eq. (9.9).

We will now obtain the threshold powers for SRS and SBS processes for a linear density profile with scale length L_n,

$$\omega_{\mathrm{p}}^2 = \omega_{\mathrm{p0}}^2 \left(1 + \frac{x}{L_n}\right) .$$

Stimulated Raman scattering

In the case $v_{gx} = 3v_{\mathrm{th}}^2 k_x/2\omega$, $v_{g1x} = c^2 k_{1x}/\omega_1$, $K' \cong \omega_{\mathrm{p0}}^2/3L_n k_x v_{\mathrm{th}}^2$, the convective amplification factor A first derived by one of the authors (CSL) in 1972

$$A = \frac{k^2 |v_0|^2 L_n}{8c^2 k_{1x}} . \tag{9.15}$$

For the backscattering process in the underdense region Eq. (9.15) gives the threshold Poynting flux (power flow density) corresponding to $A = 1$,

$$P_{\mathrm{th}} \cong \frac{m n_{\mathrm{cr}} c^3}{(\omega_0 L_n/c)} , \tag{9.16}$$

where $n_{\mathrm{cr}} = m\omega_0^2/4\pi e^2$ is the critical density.

Numerous laser plasma experiments have confirmed the enhancement of Raman reflectivity with the scale length L_n of the underdense plasma. Figure 9.1 shows the time-integrated fraction of Raman reflected laser light as a function of density scale length. The boxes denote the range of the data from numerous experiments using 0.53μ laser light at $I \gtrsim 10^{15}$ W/cm^2. Long scale length plasmas ($L_n \sim 10^3\lambda_0$) are produced by irradiating exploding foil targets, where appropriate plasma conditions are only maintained over a portion of the laser pulse. Hence, the time-integrated Raman reflectivity of 15% may in fact correspond to instantaneous reflectivity of 20–30%. Such a high value of reflectivity is within a factor of two or three of the maximum possible reflectivity by the Manley-Rowe conditions.

Stimulated Brillouin scattering

For $\omega \ll \omega_{\mathrm{pi}}$ the wave number of the ion acoustic wave is independent of electron density and K' in a stimulated Brillouin scattering process due to

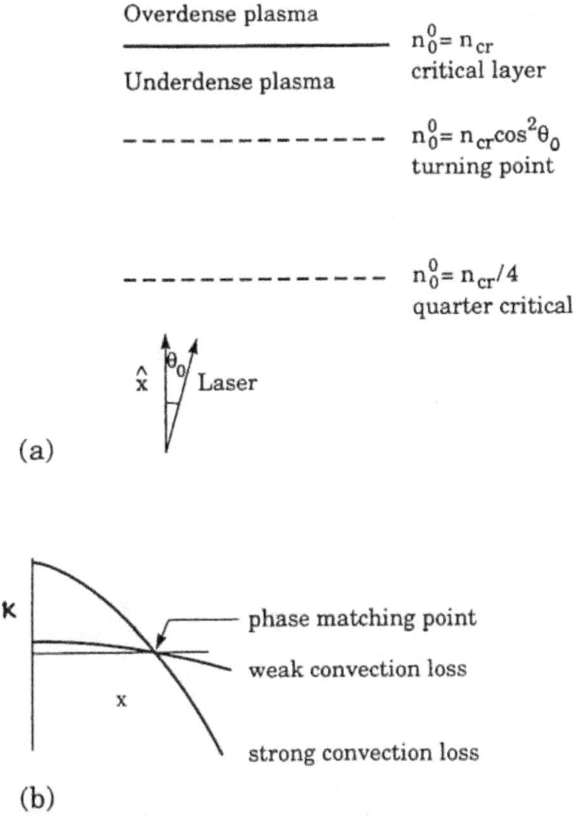

Fig. 9.1. (a) Various regions of an inhomogeneous plasma. (b) Wave number mismatch $K = k - (k_0 + k_1)$ as a function of x for three-wave decay in an inhomogeneous plasma.

density gradient is rather small, leading to large amplification. However, as the sound speed $c_s = (T_c/M)^{1/2}$, temperature inhomogeneity and the nonuniformity in plasma flow velocity introduce a x dependence of k, enhancing the value K'. In a laser produced plasma the plasma has an outward flow with a nonuniform flow velocity $\hat{x}u(x)$. The scale length of variation u is shorter than that of T_e, hence we consider the nonuniformity arising only through $u(x)$. The plasma flow modifies the ion susceptibility

$$\chi_i = -\frac{\omega_{pi}^2}{(\omega - k_x u(x))^2} \tag{9.17}$$

giving the modified dispersion relation for ion acoustic waves

$$\omega = \frac{kc_s}{(1 + k^2c_s^2/\omega_{pi}^2)^{1/2}} + k_x u(x)$$
$$\cong kc_s + k_x u \ . \tag{9.18}$$

Then $v_{gx} = u + (k_x/k)c_s$, $K' \simeq (dk_x/dx) \simeq \{k_x u/[u + (k_x/k)c_s]\}1/L_u$ and the convective amplification factor turns out to be

$$A = \frac{\gamma_M^2 L_u k_1}{cuk_x k_{1x}} \ , \tag{9.19}$$

where γ_M is given by Eq. (7.57) and L_u is the scale length of u variation. For backscatter

$$A = \frac{|v_0|^2}{8v_{th}^2}\frac{\omega_p^2}{\omega_0^2}\left(\frac{\omega_0}{c}L_u\right)\frac{c_s}{u}, \tag{9.20}$$

$$P_{th} = 4\frac{n_{cr}}{n_0^0}\frac{u/c_s}{L_u\omega_0/c}mv_{th}^2 n_{cr}c \ . \tag{9.21}$$

9.2. Raman Side Scattering

The above treatment is valid as long as $(d/dx)\ln k_x \ll k_x$ holds for both the daughter waves. When the angle between $\mathbf{k}_1$ and ∇n_0^0 approaches $\pi/2$ or the interaction region is close to the quarter critical layer, $k_x \to 0$ and this approximation fails. Then one must solve the fluid and Maxwell equations without replacing $d\phi/dx$ by ik_x, $\partial E_1/\partial x$ by $-ik_{1x}$. The same result can be obtained by replacing $k_x^2\phi$ by $-\partial^2\phi/\partial x^2$ and $k_{1x}^2 E_1$ by $-\partial^2 E_1/\partial x^2$ in the coupled mode equations (7.43) and (7.46)

$$\frac{d^2\phi}{dx^2} + \left[\frac{\omega^2 - \omega_p^2}{v_{th}^2} - k_z^2 + \frac{2i\omega\Gamma}{v_{th}^2}\right]\phi = |v_0|\frac{E_1\omega^2 e^{ik_0x}}{2\omega_1 v_{th}^2} \ , \tag{9.22}$$

$$\frac{d^2 E_1}{dx^2} + \left[\frac{\omega_1^2 - \omega_p^2}{c^2} - k_{1z}^2\right]E_1 = |v_0|\frac{\omega_1 e^{-ik_0x}}{2c^2}\left(k_z^2\phi - \frac{d^2\phi}{dx^2}\right) \ . \tag{9.23}$$

Our interest is in the neighborhood of the turning point of E_1, where the second term in Eq. (9.23) vanishes, i.e., $k_{1x}^2 \approx 0$. For phase matching $k_x \cong k_{0x}$. To solve this coupled set analytically we neglect thermal effects on the Langmuir wave, i.e., $\omega^2 - \omega_p^2 > k^2 v_{th}^2, 2i\omega\Gamma$, consider $\omega_p^2 = \omega_{p0}^2(1+x/L_n)$ and introduce new variables $\xi = x/\lambda_{em}$, $\lambda_{em} = \left(c^2 L_n/\omega_p^2\right)^{1/3}$, obtaining from (9.22) (9.23)

$$\frac{d^2 E_1}{d\xi^2} - \left[\xi - D - \frac{\lambda}{\xi - B}\right] E_1 = 0 , \tag{9.24}$$

where

$$D = \left[\frac{\omega_1^2 - \omega_{p0}^2}{c^2} - k_{1z}^2\right] \lambda_{em}^2$$

$$B = \left(\frac{\omega^2}{\omega_{p0}^2} - 1\right) L_n/\lambda_{em}$$

$$\lambda = L_n \lambda_{em} \frac{|v_0|^2 k^2}{4c^2} .$$

Changing the variable further to

$$\xi = \xi - \frac{D+B}{2} ,$$

Eq. (9.24) becomes

$$\frac{d^2 E_1}{d\xi^2} + \frac{\alpha^2 + \lambda - \xi^2}{\alpha + \xi} E_1 = 0 , \tag{9.25}$$

where $\alpha = (D - B)/2$.

This equation has two turning points at

$$\xi_T = \pm(\alpha^2 + \lambda)^{1/2}$$

As long as $|\xi_T| \ll \alpha$, as must be verified *a posteriori*, Eq. (9.25) offers parabolic cylinder function solutions and the eigenvalue equation is

$$\frac{\alpha^2 + \lambda}{2\sqrt{\alpha}} = n + \frac{1}{2}, \qquad n = 0, 1, 2, \ldots . \tag{9.26}$$

Solving (9.26) iteratively for large $\lambda \gg 1$,

$$\alpha = -i\sqrt{\lambda}\left[1 + \left(n + \frac{1}{2}(-\lambda)^{-3/4}\right)\right] . \tag{9.27}$$

The condition $|\xi_T| \ll \alpha$ reduces to $\lambda^{1/4} \gg (2n+1)$ or

$$\frac{|v_0|}{c} \gg (2n+1)^2 \frac{2}{k\sqrt{L_n \lambda_{\rm em}}} = (2n+1)^2 \frac{2}{(kc/\omega)^{1/3}(kL_n)^{2/3}} \tag{9.28}$$

The growth note on solving $\alpha = (D-B)/2$ turns out to be

$$\gamma = \sqrt{\lambda}\left(\frac{c}{L_n \omega_{\rm p0}}\right)^{2/3} \frac{1}{(|\omega_1|/\omega_{\rm p0}+1)} \left[1 - \frac{\left(n+\frac{1}{2}\right)\left(\frac{\omega_{\rm p}}{kc}\right)^{1/2}}{\sqrt{2}\left(\frac{v_0}{c}\right)^{3/2} kL_n}\right]$$

$$= \frac{k|v_0|}{2\omega_0}\left[1 - \frac{\left(n+\frac{1}{2}\right)\left(\frac{\omega_{\rm p}}{kc}\right)^{1/2}}{\sqrt{2}\left(\frac{v_0}{c}\right)^{3/2} kL_n}\right] . \tag{9.29}$$

The threshold is approximately given by the condition $|\xi_T| < \alpha$ by Eq. (9.28) or $\gamma > 0$ in (9.29).

9.3. Brillouin Side Scattering

When the sideband propagates nearly transverse to the density gradient, having a turning point in the interaction region, one cannot employ a WKB approximation for it. The wave equations for the sideband and the acoustic wave are

$$\frac{d^2E_1}{dx^2} + \left[\frac{\omega_1^2}{c^2}\left(1 - \frac{\omega_{\rm p}^2}{\omega_1^2}\right) - k_{1z}^2\right] E_1 = \frac{i\omega_1}{2}\frac{\omega_{\rm pi}^2}{c_s^2}\frac{v_0^*}{c^2}\phi , \tag{9.30}$$

$$2ik_x \frac{d\phi}{dx} = \frac{\omega^2}{c_s^2}\frac{v_0}{2i\omega_1} E_1 . \tag{9.31}$$

Let $\omega_{\rm p}^2 = \omega_{\rm p0}^2(1 + x/L_n)$, $\omega_{\rm p0}^2 = \omega_1^2 - k_{1z}^2 c^2$, $x' = x/\lambda_{\rm em} = (c^2 L_n/\omega_{\rm p0}^2)^{1/3}$. Hence,

$$\frac{d^2E_1}{dx'^2} - x'E_1 = -\frac{i\omega_0}{2}\frac{\omega_{\rm pi}^2}{c_s^2}\frac{v_0^*}{c^2}\lambda_{\rm em}^2 \phi , \tag{9.32}$$

$$\frac{d\phi}{dx'} = \frac{\omega^2 v_0 \lambda_{\rm em}}{4\omega_0 k_x c_s^2} E_1 . \tag{9.33}$$

Differentiating Eq. (9.32) w.r.t. x and using (9.33) we get

$$\frac{d^3E_1}{dx'^3} - \frac{d}{dx'}(x'E_1) + iGE_1 = \delta(x') \tag{9.34}$$

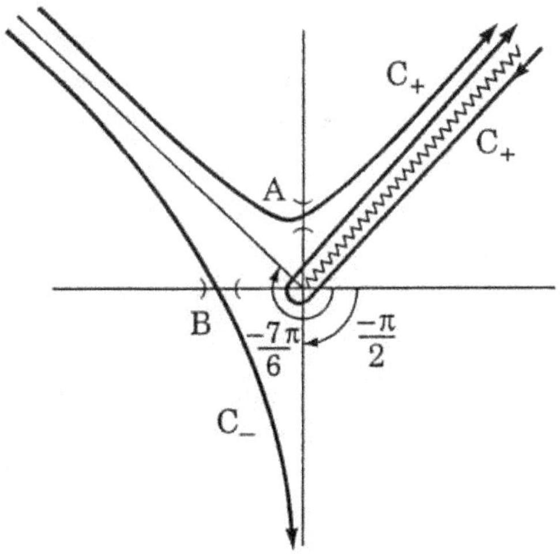

Fig. 9.2. Contours of integration in the complex k-plane. Saddle point A is for $x' > 0$: $k = i\sqrt{x'}$; and B is for $x' < 0$: $k = -\sqrt{-x'}$.

where $G = (\omega^2|v_0|^2/8c_s^2c^2)(\omega_{pi}^2\lambda_{em}^3/c_s^2k_x)$, $k_x \simeq k_0$, a delta function source has been introduced to examine convective amplification. First we solve the homogeneous equation to obtain solutions for $x' < 0$ and $x' > 0$, then match E_1 and dE_1/dx at $x' = 0$, and apply the jump condition on d^2E_1/dx'^2 which, on integrating Eq. (9.34) across $x' = 0$ turns out to be

$$\left.\frac{d^2E_1}{dx'^2}\right|_{0-}^{0+} = 1 \tag{9.35}$$

To solve the homogeneous equation we express

$$E_1 = \int_c E_{1k}e^{ikx'}dk$$

Then Eq. (9.34) leads to

$$\frac{dE_{1k}}{dk} + \left(-ik^2 + \frac{iG}{k}\right)E_{1k} = 0 \tag{9.36}$$

and

$$E_1 = A\int_c e^{i\left(\frac{k^3}{3} - G\ln k + kx'\right)}dk \tag{9.37}$$

where contours are to be so chosen to give proper behavior at $x' = \pm\infty$. Since Eq. (9.34) is a third order differential equation it must have three independent solutions, hence three contours. One must realize that for $x' \to -\infty$ one should have only an outward going electromagnetic wave whereas for $x' \to \infty$ one must have an outward ion acoustic wave and a decaying (evanescent) electromagnetic wave.

For large k the integrand dominated by $e^{ik^3/3}$ falls off most rapidly along three radius vectors (straight lines running from 0 to ∞) along $k = \rho e^{i\theta}$ or $k^3 = \rho^3(\cos 3\theta + i \sin 3\theta)$ such that $\sin 3\theta = 1$ (cf. Fig. 9.2) Further there is a branch point at $k = 0$. One may choose a branch cut along one of these lines, say $\theta = \pi/6$. Then the other two lines along which the integrand falls most rapidly are $\theta = -\pi/2, -7\pi/6$. To determine the appropriate contours we evaluate the saddle points and obtain asymptotic behavior. The saddle points are obtained by setting $(d/dk)\,(k^3/3 - G \ln k + kx') = 0$. For large x' they are $k = \pm i\sqrt{x'}$. Recalling the method of steepest descent to evaluate the integrals of the kind

$$I(\alpha) = \int_c e^{\alpha f(z)} dz \approx \left(\frac{2\pi}{\alpha\rho}\right)^{1/2} e^{\alpha f(z_0)} e^{i\phi}$$

(where z_0 is the saddle point at which $df/dz = 0$, $\rho e^{i\theta} = d^2 f/dz^2$ at z_0, $z - z_0 = s e^{i\phi}$), we have

$$\begin{aligned} E_1(x' \to \infty) &\sim e^{\pm\frac{2}{3}x'^{3/2}} \\ E_1(x' \to -\infty) &\sim e^{\pm i\frac{2}{3}(-x')^{3/2}}\,, \end{aligned} \tag{9.38}$$

representing usual electromagnetic waves. The contribution from the neighborhood of $k \sim \frac{G}{x'}$ (≈ 0 for large x') gives the ion mode

$$E_1(x \to \infty) = \left(\frac{G}{x'}\right)^{1-iG} \tag{9.39}$$

as seen by equating the last two terms of Eq. (9.37) and realizing that $\phi(x \to \infty) = \text{const}$. Thus the appropriate contour, which asymptotically gives the ion wave, starts from ∞ to origin along the $\theta = \pi/6$ line, circles the origin and goes back to infinity along the same line. The contours giving well behaved electromagnetic waves for $x' \gtrless 0$ are also shown in Fig. 9.2. For large x' they should pass through the appropriate saddle points. Thus we may write

$$E_{1I} \equiv E_1(x' > 0) = g \int_{c+} f(k)dk + h \int_{c'_+} f(k)dk$$

$$E_{1II} \equiv E_1(x' < 0) = j \int_{c-} f(k)dk \; ; \tag{9.40}$$

$$\int_{c+} = \int_{\theta=\frac{\pi}{6}} - \int_{\theta=-\frac{11}{6}\pi}$$

$$\int_{c'_+} = \int_{\theta=-\frac{11}{6}\pi} - \int_{\theta=-\frac{7\pi}{6}}$$

$$\int_{c-} = \int_{\theta=-\frac{11\pi}{6}} - \int_{\theta=-\frac{7\pi}{6}}$$

where $f(k) = e^{i(k^3/3 - G\ln k + kx)}$ and $\int_{\theta=\frac{\pi}{6}}$ implies the integral from 0 to ∞ along the radius vector $\theta = \pi/6$. First and second derivatives of E_1 can be obtained from Eqs. (9.40) by replacing $f(k)$ by $ikf(k)$ and $-k^2 f(k)$ respectively. At $x = 0$ these integrals are expressible in terms of gamma functions, e.g.,

$$\int_{\phi} e^{ik^3/3} k^{-iG} dk$$

$$= \int_0^{\infty} e^{-|k|^3/3} |k|^{-iG} (e^{i\phi})^{-iG+1} d|k|$$

$$= e^{\phi(G+i)} \int_0^{\infty} e^{-t} t^{-(iG+2)/3} dt [3^{-(2+iG)/3}]$$

$$= e^{\phi(G+i)} 3^{-(2+iG)/3} \Gamma\left[\frac{(1 \dot{-} iG)}{3}\right] \tag{9.41}$$

Demanding the continuity of $E_1, dE_1/dx'$ at $x' = 0$ and the jump in $d^2E/dx'^2|_{0-}^{0+} = 1$ we can evaluate the constants. Retaining the highest power of $e^{\pi G}$ we obtain

$$E_1(0) \simeq e^{\pi G} \left[e^{-4\pi i/6} \frac{(1 + e^{-\pi i/3})}{(1 + e^{+i\pi/3})} \frac{3^{-iG}}{\Gamma\left(\frac{1}{3} - \frac{iG}{3}\right)} \right] \tag{9.42}$$

Comparing this value of $E_1(0)$ with the one obtained when $G = 0$ (let the latter be $E_{10}(0)$) we obtain the amplification

$$\left|\frac{E_1(0)}{E_{10}(0)}\right|^2 \sim e^{2\pi G} .$$

Thus the convective threshold for the side scattering instability is $G \geq 1$, i.e.,

$$\frac{v_0^2}{4v_e^2}\frac{\omega_0 L_n}{c} \geq 1 \tag{9.43}$$

The same condition could be deduced also by evaluating the convective amplification factor

$$\begin{aligned} A &\equiv \frac{\gamma_0^2}{v_{gx}v_{g1x}\frac{d}{dx}(k_x - k_{1x} - k_{0k})} \\ &\approx \frac{\gamma_0^2}{c_s\frac{d\omega_1}{dx}} \approx \frac{\gamma_0^2 2\omega_0 L_n}{c_s\gamma_p^2} \\ &\approx \frac{v_0^2}{4v_e^2}\frac{\omega_0 L_n}{c} \gtrsim 1 . \end{aligned} \tag{9.44}$$

Thus the major factor responsible for the enhancement of sidescattering is the slowing down of the group velocity of the light wave to

$$v_{g1x} \sim c\frac{\lambda_0}{\lambda_{\rm em}} = c\frac{c/\omega_0}{(c^2L_n/\omega_{\rm p0}^2)^{1/3}} \sim c\left(\frac{c}{L_n\omega_{\rm p0}}\right)^{1/3}\frac{\omega_{\rm p0}}{\omega_0} .$$

9.4. Scattering Off a Heavily Damped Ion Acoustic Mode

In cases where the ion temperature is comparable to the electron temperature or else electron-ion and electron-neutral collisions are important, the ion acoustic mode is heavily damped, and strongly localized near the interaction region. The strong localization of the instable region provides partial reflections of the reflected wave leading to absolute instability. Retaining $\partial/\partial t$ and the damping term, the equation governing ϕ can be written as

$$k_x\frac{\partial\phi}{\partial x} + \frac{\omega}{c_s^2}\frac{\partial}{\partial t}\phi + \Gamma\frac{\omega}{c_s^2}\phi = -\frac{\omega^2}{4c_s^2}\frac{v_0}{\omega_1 k_x}E_1 . \tag{9.45}$$

Taking the time variation of ϕ as e^{pt}, replacing $\omega_1 \to \omega_1 + ip$ in Eq. (9.30), we obtain, for $(\Gamma + p)\phi > \frac{k_x}{\omega}c_s^2(\partial\phi/\partial x)$,

$$\frac{d^2E_1}{dx'^2} - x'E_1 - i\left[p - \frac{\gamma_M^2}{p+\Gamma}\right]\frac{\omega_0}{c^2}\lambda_{\rm em}^2 E_1 = 0 \tag{9.46}$$

where γ_M is given by Eq. (7.57). This is an eigenvalue equation for p, subject to the boundary conditions that E_1 be evanescent at large $+x'$ and outgoing at negative large x'. At large x' E_1 behaves asymptotically

$$E_1 \sim e^{\mp i}\int(-x' - i\alpha)^{1/2}dx'$$
$$= e^{\pm}\left[i\frac{2}{3}(-x')^{3/2} + \alpha(-x')^{1/2}\right] \tag{9.47}$$

where upper and lower signs refer to $x \to -\infty$ and $+\infty$ and $\alpha = p - \gamma_M^2/(p + \Gamma)$. For solution to be well behaved α must be negative, i.e.,

$$p \leq \frac{\gamma_M^2}{p + \Gamma} .$$

The largest growth rate is obtained when the equality sign holds

$$\gamma^2 + \gamma\Gamma = \gamma_M^2$$
$$\gamma = \frac{1}{2}\left[-\Gamma + \sqrt{\Gamma^2 + 4\gamma_M^2}\right] . \tag{9.48}$$

For values of $p \to 0$, Eq. (9.46) gives convective amplification. To evaluate the amplification factor we introduce a delta function source term in Eq. (9.46)

$$\frac{d^2E_1}{dx'^2} - x'E_1 + i\frac{\gamma_M^2}{\Gamma}\frac{\lambda_{\rm em}^2\omega_0}{c^2}E_1 = \delta(x') . \tag{9.49}$$

First we solve the homogeneous equation. Taking its Fourier transform

$$\frac{d}{dk}E_{1k} + (ik^2 + \alpha_0)E_{1k} = 0$$
$$\alpha_0 \equiv \frac{\gamma_M^2}{\Gamma}\frac{\lambda_{\rm em}^2\omega_0}{c^2} \tag{9.50}$$

we obtain $E_{1k} = Ae^{-(ik^3/3 + \alpha_0 k)}$,

$$E_1 = \frac{A}{2\pi}\int_c e^{-(i\frac{k^3}{3} + \alpha_0 k - ikx)}dk \tag{9.51}$$

To determine the appropriate contours we obtain the saddle points and study the asymptotic behavior at large x. The saddle points are given by

$$k^2 - i\alpha_0 - x = 0$$

which for large x are

$$k = \pm x^{1/2} \tag{9.52}$$

giving asymptotic behavior of E_1. Following the procedure of Sec. 9.3 the appropriate contours are $c_+ = c_3 - c_1$ and $c_- = c_2 - c_1$ where c_1, c_2, c_3 are the lines $\theta = \pi/6$, $5\pi/6$ and $9\pi/6$ of steepest descent. Thus

$$\begin{aligned} E_{1\mathrm{I}} &\equiv E_1(x > 0) = g_+ \int_{c+} f(k)dk \\ E_{1\mathrm{II}} &\equiv E_1(x < 0) = g_- \int_{c-} f(k)dk \,, \end{aligned} \tag{9.53}$$

where $f(k)$ is the integrand of Eq. (9.51). Demanding the continuity of E_1 at $x' = 0$ and the jump in $\frac{dE_1}{dx'}$ at $x' = 0$ (obtained from Eq. (9.49)) to be

$$\left.\frac{dE_1}{dx'}\right|_{0^-}^{0^+} = 1$$

one obtains the convective amplification factor

$$\frac{|E_1|^2_{\alpha_0 \neq 0}}{|E_1|^2_{\alpha_0 = 0}} = e^{\frac{2\sqrt{2}}{3}\alpha_0^{3/2}} \,. \tag{9.54}$$

The threshold condition is

$$\frac{\alpha_0^{3/2}}{3} \gtrsim 1 \,. \tag{9.55}$$

9.5. Decay Instability

An electromagnetic pump wave propagating along the density gradient has an Airy function field profile at the critical layer with a scale length $\lambda_{\mathrm{em}} = (c^2 L_n/\omega^2)^{1/3}$. It can parametrically drive the decay instability producing ion acoustic and Langmuir modes of short wavelength, $k\lambda_{\mathrm{em}} \gg 1$. Since the wave vector of the maximally growing mode (cf. Eq. (7.82)) is aligned along the field of the pump (i.e., $\mathbf{k} \perp \nabla n_0^0$), the Langmuir wave spends a long time near the phase matching point and possesses a turning point in its neighborhood. We examine the problem analytically assuming the ion acoustic mode to be strongly damped (i.e., $\varepsilon \neq 0$). Then Eq. (7.78) can be cast into the form of a wave equation for ϕ_1 by replacing $\mathbf{k}_1 = \hat{y}k_1 - i\hat{x}\,\partial/\partial x$:

$$\frac{d^2\phi_1}{dx^2} - \frac{x}{\lambda_{\mathrm{es}}^3}\phi_1 + \alpha\phi_1 = 0 \tag{9.56}$$

where $\lambda_{es} = (3v_{th}^2 L_n/2\omega_1^2)^{1/3}$, $\alpha = (2\chi_e\chi_i/3\varepsilon)(k^2|v_0|^2/4v_{th}^2)$ and we have assumed a linear density profile $\omega_p^2 = \omega_{p0}^2(1 + X/L_n)$ with $\omega_{p0}^2 = \omega_1^2 - 3k^2v_{th}^2/2$. Since $\chi_e \simeq \omega_{pi}^2/k^2c_s^2$, $\chi_i \simeq -\omega_{pi}^2/\omega^2$,

$$\varepsilon \simeq (\omega_{pi}^2/k^2c_s^2w^2)(\omega^2 - k^2c_s^2 + 2i\Gamma\omega) \simeq 2i\Gamma\frac{\omega_{pi}^2}{k^2c_s^2\omega}\,.$$

Eq. (9.56) can be cast into the form of Eq. (9.46)

$$\frac{d^2\phi_1}{dx'^2} - x'\phi_1 + iG^2\phi_1 = 0 \tag{9.57}$$

where $G^2 = (\omega_{pi}^2/3\omega\Gamma)(k^2|v_0|^2\lambda_{es}^2/4v_{th}^2)$, $x' = x/\lambda_{es}$. Following the treatment of Sec. 9.4, the convective amplification factor can be written as (cf. Eq. (9.54)) e^A where $A = \frac{\sqrt{2}}{3}G^3$. The threshold is thus given by

$$\frac{|v_0|}{v_{th}} \gtrsim \frac{(\omega\Gamma)^{1/2}}{\omega_{pi}}\frac{1}{k\lambda_{es}}\,. \tag{9.58}$$

Oblique pump

An electromagnetic wave obliquely incident on an inhomogeneous plasma encounters a turning point at $\omega_p = \omega_0 \cos\theta_0$ where θ_0 is the angle $\mathbf{k}_0$ makes with $\hat{x}$ at the entry point. When the separation between the turning point and the critical layer is small, the field of the *p*-polarized light tunnels to the critical layer, acquiring a large value of E_{0x}. Following the references, E_{0x} may be approximated as

$$E_{0x} = \frac{E_{in}\Phi}{2\pi(k_0L_n)^{1/2}}\frac{L_n}{x + i\Delta} \tag{9.59}$$

where E_{in} is the incident wave amplitude, $k_0 = \omega_0/c$, Φ is a function of $(\omega_0L_n/c)^{1/3}\sin\theta_0$ and is of the order of unity for a narrow range of θ_0, and Δ is the width of the resonance $\Delta \sim \nu L_n/\omega_p$ or $(L_nv_{th}^2/\omega_p^2)^{1/3}$, depending on whether collisional damping or convection losses are predominant. Such a spatially localized pump may give rise to temporally growing decay instability. Now, since $\mathbf{E}_0 \| \hat{x}$ the $\mathbf{k}$ vector of the maximally growing decay waves is along the density gradient. We may employ a WKB approximation for both the ion acoustic and Langmuir waves. From Eqs. (7.75) and (7.78) we thus write, taking $k \to k - i\frac{\partial}{\partial x'}$, $\omega \to \omega + ip$

$$\frac{d\phi}{dx}+\frac{p}{c_s}\phi=\frac{ikv_{00}\omega}{4c_s\omega_1}\phi_1 e^{-iK'x^2/2}\frac{\Delta}{x+i\Delta}\,, \tag{9.60}$$

$$\frac{d\phi_1}{dx}+\frac{2}{3}\frac{\omega_1 p}{k_1 v_{\text{th}}^2}\phi_1=\frac{iv_{00}^*\omega_1}{6v_{\text{th}}^2}\frac{\omega_{\text{pi}}^2}{\omega^2}\phi e^{+iK'x^2/2}\frac{\Delta}{x-i\Delta}\,, \tag{9.61}$$

where $K' = d(k-k_1)/dx \simeq -\omega_{\text{p}}^2/3v_{\text{th}}^2 k_1 L_n$, $v_{00} = eE_{\text{in}}\Phi L_n/[2\pi i\Delta(k_0L_n)^{1/2}m\omega w]$. Multiplying Eq. (9.61) by $(x-i\Delta)\exp(-iK'x^2/2)$, operating by $(p/c_s+\partial/\partial_x)$ and using Eq. (9.60) we obtain an equation for ϕ_1. Introducing a function

$$\psi_1=\phi_1\exp\left[\frac{1}{2}\int\left(\frac{p}{c_s}-\frac{2}{3}\frac{p\omega_0}{kv_{\text{th}}^2}-iK'x\right)dx\right]\,, \tag{9.62}$$

and assuming $\Delta p/c_s > 1$, $kv_{\text{th}} > \omega_{\text{pi}}$ the equation for ϕ_1 yields

$$\frac{d^2\psi_1}{dx^2}+\left[\frac{\alpha^2}{x^2+\Delta^2}-\frac{1}{4}\left(\frac{p}{c_s}-iK'(x)^2\right)\right]\psi_1=0 \tag{9.63}$$

where

$$\alpha^2=\frac{\omega_{\text{pi}}^2\Delta^2}{24c_s^2}\frac{|v_{00}|^2}{v_{\text{th}}^2}\,.$$

For $p/c_s > Kx_T$, ψ_1 has turning points at $x_T=\pm(4c_s^2\alpha^2/p^2-\Delta^2)^{1/2}$. Further, when $x_T \ll \Delta$, Eq. (9.63) goes over to a harmonic oscillator equation with the eigenvalue condition

$$\left(\alpha^2-\frac{p^2\Delta^2}{4c_s^2}\right)\frac{2c_s}{p\Delta}=2n+1;\qquad n=0,1,2,\ldots \tag{9.64}$$

Maximum growth occurs for $n=0$. Substituting the value of α and taking $\Delta\sim(v_{\text{th}}^2L_n/\omega_0^2)^{1/3}$, the growth rate is approximately given by

$$p=\frac{\omega_{\text{pi}}}{2\pi\sqrt{6}}\left(\frac{L_n\omega_{\text{p}}}{v_{\text{th}}}\right)\frac{eE_{\text{in}}\Phi}{(k_0L_n)^{1/2}m\omega_0 v_{\text{th}}}\,. \tag{9.65}$$

In cases where the turning points are far apart ($x_T \gtrsim \Delta$) Eq. (9.63) can be solved numerically to obtain eigenvalues.

9.6. Oscillating Two-Stream Instability

We have seen that OTSI grows when the frequency mismatch $\Delta\equiv\omega_0-\omega_{\text{p}}\left(1+3k^2v_{\text{th}}^2/2\omega_{\text{p}}^2\right)^{1/2}<0$. For $\mathbf{E}_0\|\hat{x}$, the growth rate maximizes to γ_{max} for $\Delta=\gamma_{\text{max}}\equiv(k_z^2/4k^2)(|v_0|^2/v_{\text{th}}^2)\omega_{\text{p}}$. In an inhomogeneous plasma with $\nabla n_0\|\hat{x}$, $k^2=k_x^2(x)+k_z^2$. As the Langmuir sidebands propagate towards the underdense region $k_x \equiv \left\{\{[\omega^2 - \omega_p^2(x)]/(3v_{th}^2/2)\} - k_z^2\right\}^{1/2}$ increases rapidly leaving Δ unchanged but diminishing γ_{max} severely. Since $\Delta = \gamma_{max}$ becomes invalid the growth rate $\gamma \ll \gamma_{max}(x)$. This strong localization of the interaction provides substantial reflections of the sidebands leading to an absolute instability. Here we follow the treatment of Liu to obtain the threshold for the onset of the absolute instability.

In order to account for the effects of sideband wave reflection at the critical layer we start with the full wave equation for the sidebands,

$$\frac{\partial^2 \phi_j}{\partial x^2} - k_z^2 \phi_j = 4\pi e \left(n_j^{L} + n_j^{NL}\right), \qquad j = 1, 2, \ldots, \tag{9.66}$$

The linear density perturbations n_j^L can be written as

$$n_j^L = \frac{1}{i\omega_j} \nabla \cdot \left[n_0^0 \left(-\frac{e\nabla\phi_j}{mi\omega_j} \right) + \frac{v_{th}^2}{i\omega_j} \nabla n_j^L \right]$$

$$\simeq \frac{e}{m\omega_j^2} \nabla \cdot (n_0^0 \nabla \phi_j) - \frac{v_{th}^2}{\omega_j^2} \nabla^4 \phi_j \frac{n_0^0 e^2}{m\omega_j^2} \,. \tag{9.67}$$

Using Eqs. (7.93) and (9.67) in Eq. (9.66) and taking $\omega_p^2 \equiv 4\pi n_0^0 e^2/m = \omega_{p0}^2(1 + x/L_n)$, $\phi_j = \Phi_j \exp[-i(\omega_j t - k_z z)]$ we obtain

$$\frac{d^2 \Phi_j}{dx^2} - k_z^2 \Phi_j = \frac{\omega_{p0}^2}{\omega_j^2} \left(1 + \frac{x}{L_n}\right) \left(\frac{d^2}{dx^2} - k_z^2\right) \Phi_j$$

$$+ \frac{\omega_{p0}^2}{\omega_j^2 L_n} \frac{d}{dx} \Phi_j - \frac{3}{2} \frac{\omega_{p0}^2 v_{th}^2}{\omega_j^4} \left(\frac{d^2}{dx^2} - k_z^2\right)^2 \phi_j$$

$$+ \frac{\omega_{p0}^2 k_z^2 |v_0|^2}{4v_{th}^2 (1 + T_i/T_e)\omega_j} \left(\frac{\Phi_1}{\omega_1} + \frac{\Phi_2}{\omega_2}\right), \tag{9.68}$$

where we have assumed, without any loss of generality $A_0 = i|A_0|$. Expressing

$$\Phi_j = \int_c \Phi_{jk_x} e^{ik_x x} dk_x \,, \tag{9.69}$$

Eq. (9.68) takes the form

$$\frac{d}{dk_x} \Phi_{jk_x} = -i \frac{\omega_j^2 L_n}{\omega_{p0}^2} \left[1 - \frac{\omega_{p0}^2}{\omega_j^2} - i \frac{\omega_{p0}^2}{\omega_j^2} \frac{k_x}{k^2 L_n} - \frac{3}{2} \frac{\omega_{p0}^2}{\omega_j^2} \frac{k^2 v_{th}^2}{\omega_j^2}\right] \Phi_{jk_x}$$

$$- \frac{i w_j |v_0|^2 L_n}{4v_{th}^2 (1 + T_i/T_e)} \frac{k_2^2}{k^2} \cdot \left(\frac{\Phi_{1k_x}}{\omega_1} + \frac{\Phi_{2k_x}}{\omega_2}\right), \tag{9.70}$$

where $k^2 = k_x^2 + k_z^2$. Near the threshold $\omega \simeq 0$, i.e., $\omega_{1,2} = \mp\omega_0$. Further, without any loss of generality we define $\omega_{p0}^2 \equiv \omega_0^2 + 3k^2 v_{th}^2/2$. Defining $\Phi_\pm = \Phi_{1k_x} \pm \Phi_{2k_x}$, Eq. (9.70) yields

$$\frac{d\Phi_+}{dk_x} = \left(iL_n \frac{k_x^2 v_{th}^2}{\omega_0^2} - \frac{k_x}{k_x^2 + k_z^2}\right)\Phi_+$$

$$\frac{d\Phi_-}{dk_x} = \left(iL_n \frac{k_x^2 v_{th}^2}{\omega_0^2} - \frac{k_x}{k_x^2 + k_z^2}\right)\Phi_- - i\frac{|v_0|^2 L_n}{2v_{th}^2(1 + T_i/T_e)} \frac{k_z^2}{k_x^2 + k_z^2}\Phi_- \tag{9.71}$$

giving

$$\begin{aligned} \Phi_+ &= A(k_x^2 + k_z^2)^{-1/2} e^{ik_x^3 \lambda_{es}^3/3} \\ \Phi_- &= B(k_x^2 + k_z^2)^{-1/2} e^{i(k_x^3 \lambda_{es}^3/3 - \beta)}\,, \end{aligned} \tag{9.72}$$

where

$$\beta = \frac{|v_0|^2}{2v_{th}^2} \frac{L_n k_z}{1 + T_i/T_e} \tan^{-1}\frac{k_x}{k_z}, \qquad \lambda_{es} = (v_{th}^2 L_n/\omega_{p0}^2)^{1/3}\,.$$

From Eqs. (9.72) we may write

$$\begin{aligned} \Phi_1 &= \frac{1}{2}\int_c (k_x^2 + k_z^2)^{-1/2} e^{i(k_x^3\lambda_{es}^2/3 + k_x x)}(A + Be^{-i\beta})dk_x \\ \Phi_2 &= \frac{1}{2}\int_c (k_x^2 + k_z^2)^{-1/2} e^{i(k_x^3\lambda_{es}^2/3 + k_x x)}(A - Be^{-i\beta})dk_x\,. \end{aligned} \tag{9.73}$$

Asymptotically at large x the integral can be evaluated by the Saddle point method. The Saddle points are $k_x = \pm(-x/\lambda_{es}^3)^{1/2}$. In the underdense region as $x \to -\infty$, $\tan^{-1}(k_x/k_z) \to \pm\pi/2$. Realizing that $\phi_1 = \Phi_1 e^{+i(\omega_0 t + k_z z)}$, $\phi_2 = \Phi_2 e^{-i(\omega_0 t - k_z z)}$ the outgoing waves at $x \to -\infty$ are

$$\Phi_{1,2} = \left(k_z^2 - \frac{x}{\lambda_{es}^3}\right)^{-1/2}\left(-\frac{x}{\lambda_{es}}\right)^{-1/4} e^{\mp\frac{i}{3}(-\frac{x}{\lambda_{es}})^3}\left[A \pm Be^{\pm i\beta_0}\right]\,, \tag{9.74}$$

where $\beta_0 = |v_0|^2 L_n k_z \pi/4v_{th}^2(1 + T_i/T_e)$. The incoming waves at $x \to -\infty$ are

$$\Phi_{1,2} = \left(k_z^2 - \frac{x}{\lambda_{es}^3}\right)^{-1/2}\left(-\frac{x}{\lambda_{es}}\right)^{-1/4} e^{\pm\frac{i}{3}(-\frac{x}{\lambda_{es}})^3}\left(A' \pm B'e^{\mp i\beta_0}\right)\,. \tag{9.75}$$

However, at $x \to -\infty$ there should be no incoming sideband wave, hence,

$$\frac{|v_0|^2 k_z L_n}{4v_{\mathrm{th}}^2(1+T_{\mathrm{i}}/T_{\mathrm{e}})} = \pi \qquad \text{or} \qquad \frac{|v_0|^2}{v_{\mathrm{th}}^2} = \frac{4(1+T_{\mathrm{i}}/T_{\mathrm{e}})}{k_x L_n}\,. \tag{9.76}$$

This gives the threshold for the onset of OTSI.

9.7. Two-Plasmon Decay

Near the quarter-critical density an electromagnetic pump wave decays into two Langmuir waves. The daughter waves may couple to the pump wave to produce scattered electromagnetic waves at $\sim 3\omega_0/2$ observed in many experiments. The damping of the plasma waves heats the plasma, producing hot electrons. Plasma inhomogeneity has a profound effect on this decay process. We study it without resorting to WKB approximation. Both the decay waves have turning points on the higher density side of the resonant interaction region and propagate at an angle to the density gradient. It turns out that the decay waves, on account of nonlinear coupling, have turning points on the other side (low density side) of the interaction region also and $2\omega_{\mathrm{p}}$ decay becomes an absolute instability.

Consider an electromagnetic pump wave $\mathbf{E}_0 = \hat{z}E_0\exp[-i(\omega_0 t - k_0 x)]$ propagating along the density gradient in a plasma with linear density profile $\omega_{\mathrm{p}}^2 = \omega_{\mathrm{p0}}^2(1 + x/L_n)$. The pump wave produces oscillatory electron velocity $\mathbf{v}_0 = e\mathbf{E}_0/mi\omega_0$ and decays into two Langmuir waves of frequencies ω and ω_1 where $\omega_1 = \omega - \omega_0$. The pump and decay waves exert a ponderomotive force on the electrons $\mathbf{F}_p = e\nabla(\phi_p + \phi_{p1})$ where ϕ_p and ϕ_{p1} are the ω and ω_1 frequency components:

$$\begin{aligned}\phi_p &= -\frac{m}{2e}\mathbf{v}_0\cdot\mathbf{v}_1\\ &\simeq \frac{v_0\cdot\nabla\phi_1}{2i\omega_1}\\ \phi_{p1} &\cong \frac{\mathbf{v}_0^*\cdot\nabla\phi}{2i\omega}\,.\end{aligned} \tag{9.77}$$

Using Eqs. (9.77) the fluid equations and the Poisson equation for the ω frequency component can be written as

$$-i\omega \mathbf{v} = \frac{e}{m}\nabla\phi - \frac{v_{\rm th}^2}{n_0^0}\nabla n + \frac{e\nabla(v_0\cdot\nabla\phi_1)}{2mi\omega_1} \tag{9.78}$$

$$-i\omega n - \nabla\cdot\left[n_0^0\left(1+\frac{x}{L_n}\right)\mathbf{v} + \frac{1}{2}n_1\mathbf{v}_0\right] = 0 \tag{9.79}$$

$$\nabla^2\phi = 4\pi e n\ . \tag{9.80}$$

A similar set can be written for the ω_1 frequency component. Introducing new functions ψ, ψ_1 such that $\mathbf{v} = \nabla\psi$, $\mathbf{v}_1 = \nabla\psi_1$, taking $\partial/\partial y = 0$, $\partial/\partial z = ik_z$ and Fourier transforming in x we obtain two coupled equations for $u \equiv k^2\psi_k$, $u_1 \equiv k_1^2\psi_{k1}$:

$$\frac{i}{L_n}\frac{du}{dk_x} - \frac{\omega^2-\omega_{\rm p0}^2-k^2v_{\rm th}^2}{\omega_{\rm p0}^2}u = +\frac{k_z v_0}{2\omega_{\rm p0}}\left(\frac{k_1^2}{k^2}+\frac{\omega}{\omega_1}\right)u_1 \tag{9.81}$$

$$\frac{i}{L_n}\frac{du_1}{dk_x} - \frac{\omega_1^2-\omega_{\rm p0}^2-k_1^2v_{\rm th}^2}{\omega_{\rm p0}^2}u_1 = +\frac{k_z v_0^*}{2\omega_{\rm p0}}\left(\frac{k^2}{k_1^2}+\frac{\omega_1}{\omega}\right)u\ , \tag{9.82}$$

where $k^2 = k_x^2 + k_z^2$, $k_1^2 = (k_x - k_0)^2 + k_z^2$. Let

$$u = W\exp\left[-i\frac{L_n}{\omega_{p0}^2}\int dk_x(\omega^2-\omega_{\rm p0}^2-k^2v_{\rm th}^2)\right]$$

$$u_1 = W_1\exp\left[-i\frac{L_n}{\omega_{\rm p0}^2}\int dk_x(\omega_1^2-\omega_{\rm p0}^2-k_1^2v_{\rm th}^2)\right]$$

and $k_x = k_0/2 + K$, $\omega = \omega_0/2 + \Delta$. Then we have

$$+\frac{i}{L_n}\frac{dW}{dK} = -\frac{k_z v_0}{\omega_{\rm p0}}\frac{k_0 K W_1}{(K+k_0/2)^2+k_z^2}e^{i\theta}\ , \tag{9.83}$$

$$\frac{i}{L_n}\frac{dW_1}{dK} = -\frac{k_z v_0^*}{\omega_{\rm p0}}\frac{k_0 K W}{(K-k_0/2)^2+k_z^2}e^{-i\theta}\ , \tag{9.84}$$

where

$$\theta = \frac{L_n}{\omega_{\rm p0}^2}\int dK\left[\omega^2-\omega_1^2-2k_0Kv_{\rm th}^2\right]$$

$$= \frac{L_n}{\omega_{\rm p0}^2}2\left(\omega_0\Delta\cdot K - k_0K^2v_{\rm th}^2\right)\ .$$

Eliminating W we obtain an equation for W_1

$$\frac{d^2W_1}{dK^2} + \left[\frac{2iL_n}{\omega_{p0}^2}(\omega_0\Delta - k_0Kv_{th}^2) + \frac{2(K-k_0/2)}{(K-k_0/2)^2+k_z^2} - \frac{1}{K}\right]$$

$$\frac{dW_1}{dK} + L_n^2K^2\frac{k_z^2|v_0|^2k_0^2W_1}{\omega_0^2\left[\left(\frac{k_0^2}{4}+k_z^2+K^2\right)^2 - k_0^2K^2\right]} = 0 \tag{9.85}$$

Transforming away the first order derivation term we get

$$\frac{d^2v}{dK^2} + F(K)v = 0 \tag{9.86}$$

with $F = F_0 + F_1 + F_2$,

$$F_0 = \frac{L_n^2}{\omega_{p0}^2}\left[\frac{k_0^2|v_0|^2k_z^2K^2}{(k_z^2+k_0^2/4+K^2)^2 - K^2k_0^2} + \frac{(k_0Kv_{th}^2 - \omega_0\Delta)^2}{\omega_{p0}^2}\right],$$

$$F_1 = -i\frac{L_n}{\omega_{p0}^2}\left[\frac{\omega_0\Delta}{K} + \frac{2(K-k_0/2)(Kk_0v_{th}^2 - \omega_0\Delta)}{k_z^2+(K-k_0/2)^2}\right], \tag{9.87}$$

$$F_2 = -\frac{3}{4K^2} + \frac{(K-k_0/2)^3 - k_z^2k_0/2}{K[K_z^2+(K-k_0/2)^2]}$$

in successive orders of L_n^{-1}. The solution of Eq. (9.86) can be obtained in the WKB approximation: $W_1 \sim F^{-1/4}\exp[\pm i\int F^{1/2}dK]$. We require the solution to be localized in k space with a finite extent of localization, which implies the localization of the Fourier transformed solution in x space because of the uncertainty principle. Such a localized solution in k-space exists if (i) $F(k)$ possesses a maximum corresponding to a potential well in K space for the equivalent Schrödinger equation, and (ii) the Bohr-Sommerfield quantization condition is satisfied,

$$\int_{K_1}^{K_2} F^{1/2}dK = \left(n+\frac{1}{2}\right)\pi \tag{9.88}$$

where K_1 and K_2 are the turning points at which $F(K) = 0$, adjacent to the maximum where $dF/dK = 0$. Equation (9.88) is the eigenvalue equation determining the eigenvalues Δ, with Im Δ giving the growth rate. We obtain an approximate solution treating L_n^{-1} as a small parameter. In the limit $L_n \to \infty$ (homogeneous plasma), condition (9.88) demands that $\int_{K_1}^{K_2} Q^{1/2}dK$ be vaninshingly small where $F \equiv L_n^2Q/\omega_{p0}^2$. This implies $K_1 = K_2 = K_0$ where K_0 is the point at which F_0 has a maximum. Thus the two turning points surrounding the maximum must coalesce, i.e.,

$$F_0(K_0) = \frac{dF_0}{dK}\bigg|_{K=K_0} = 0 \,. \tag{9.89}$$

Equation (9.89) determines the location of K_0 and the lowest order eigenvalue. This condition for the coalescence of the roots with Im $\Delta > 0$ is essentially the condition for absolute instability in a homogeneous plasma. Alternatively the condition for absolute instability can be obtained by evaluating the integral $\int e^{-i(\omega(\mathbf{k})t - k_x x)} dk_x = e^{ik_0 x/2} \int e^{-i[\omega(K,k_z)t - iKx]} dK$ for the behavior of the wave packet at long time. One may conveniently employ the Saddle point method of integration. Let $q = K/k_z$. The eigen frequency from Eq. (9.87) by setting $F_0 = 0$ turns out to be

$$\omega(k_y, K) \equiv \frac{\omega_0}{2} + \Delta = \frac{\omega_0}{2} + \left[q + i\lambda\frac{q}{1+q^2}\right]\Omega \tag{9.90}$$

where $\lambda = |v_0|\omega_{p0}/k_z v_{\rm th}^2$, $\Omega = k_0 k_z v_{\rm th}^2/\omega_0$. The Saddle point is determined by $\partial\omega/\partial q = 0$, or $1 + i\lambda(1-q^2)/(1+q^2)^2 = 0$ which gives

$$q_s^2 = -\left(1 - \frac{i\lambda}{2}\right) - i\left(\frac{\lambda^2}{4} + 2i\lambda\right)^{1/2} \tag{9.91}$$

where the negative square root is picked so that $q = 1$ for $\lambda \gg 1$. Setting $q = q_s$ in Eq. (9.90), we find the value of the eigen frequency at the Saddle point

$$\Delta = q_s\left[1 + \frac{i}{2}\left\{\frac{\lambda}{2} + \left(\frac{\lambda^2}{4} + 2i\lambda\right)^{1/2}\right\}\Omega\right] \,. \tag{9.92}$$

For $\lambda \gg 1$ it gives the growth rate of the absolute instability

$$\gamma_0 = k_0|v_0|/4 \,. \tag{9.93}$$

To the next order we include F_1. Letting $K = K_0 + K_1$, $\Delta = \Delta_0 + \Delta_1$ and expanding F,

$$F = F_0(K_0, \Delta_0) + \frac{1}{2}\frac{\partial^2 F_0}{\partial K^2}\bigg|_{K_0,\Delta_0} K_1^2 + \frac{\partial F}{\partial \Delta}\bigg|_{K_0,\Delta_0} \Delta_1 + \dot{F}_1(K_0, \Delta_0)$$

(where $F_0 = \partial F_0/\partial K_0 = 0$) we rewrite Eq. (9.86) as

$$\frac{d^2 v}{dK^2} + \left[\frac{\partial F_0}{\partial K}\bigg|_{K_0,\Delta_0} \Delta_1 + F_1(K_0\Delta_0) + \frac{1}{2}\frac{\partial^2 F_0}{\partial K^2}\bigg|_{K_0,\Delta_0} K_1^2\right] \,. \tag{9.94}$$

This is the standard Weber equation with a potential well for $\partial^2 F_0/\partial K^2|_{K_0,\Delta_0} < 0$, giving

$$\Delta_1 = \left(n+\frac{1}{2}\right)\frac{(-2\partial^2 F_0/\partial K^2)^{1/2}}{\partial F_0/\partial\Delta} - \frac{F_1}{\partial F_0/\partial\Delta}$$

$$\cong -i\left(n+\frac{1}{2}\right)\frac{\omega_{p0}}{2k_z L_n} - \frac{v_{th}^2}{2|v_{th}|L_n} . \tag{9.95}$$

The growth rate including the effect of plasma inhomogeneity is therefore

$$\gamma = \mathrm{Im}(\Delta_0+\Delta_1) = \frac{k_0|v_0|}{4} - \left(n+\frac{1}{2}\right)\frac{\omega_{p0}}{2k_z L_n} . \tag{9.96}$$

The threshold condition is thus

$$\sqrt{3}k_z L_n \frac{v_0}{c} \gtrsim 1 , \tag{9.97}$$

where $k_0 \sim w_{p0}\sqrt{3}/c$ has been used. For more details one is referred to Refs. 8–10. Many numerical simulations and experiments on the two-plasmon decay have been performed and the resulting plasma waves and $(3/2)\omega_0$ scattered light studied.[11]

References

1. C. S. Liu, in *Advances in Plasma Physics*, eds. A. Simon and W. B. Thompson, (Wiley, 1976) **6**, 121.
2. M. N. Rosenbluth, *Phys. Rev. Lett.* **29**, 565 (1972); M. N. Rosenbluth, R. B. White and C. S. Liu, *Phys. Rev. Lett.* **31**, 697 (1973); R. B. White, C. S. Liu and M. N. Rosenbluth, *Phys. Rev. Lett.* **31**, 520 (1973).
3. M. Porkolab, V. Arunasalam, N. C. Luhmann Jr., and J. Schmitt, *Nucl. Fusion* **16**, 269 (1976).
4. E. M. Campbell, "Dependence of laser-plasma interaction physics on laser wavelength and plasma scalelength" in *Radiation in Plasmas* Vol. 1, ed. B. McNamara (World Scientific, Singapore, 1984) 579.
5. A. Simon, W. Seka, L. M. Goldman,, and R. W. Short, *Phys. Fluids* **29**, 1704 (1986).
6. H. C. Barr and F. F. Chen, *Phys. Fluids* **30**, 1180 (1987).
7. C. S. Liu and V. K. Tripathi, *Phys. Fluids* **29**, 4188 (1986).
8. C. S. Liu and M. N. Rosenbluth, *Phys. Fluids* **19**, 967 (1976).
9. Y. C. Lee and P. K. Kaw, *Phys. Rev. Lett.* **32**, 135 (1974).
10. A. Simon, R. W. Shurt, E. A. Williams and T. Dewandre, *Phys. Fluids* **26**, 3107 (1983).
11. L. M. Goldman, W. Seka, K. Tanaka, R. Short and A. Simon, *Con. J. Phys.* **64**, 969 (1986).

CHAPTER 10

EPILOGUE

In previous chapters, we have introduced the physics of basic processes of collective interactions of electromagnetic waves with plasma and electron beams, presenting simple physical examples wherever possible to elucidate the underlying physical mechanisms and to provide the necessary mathematical analysis.

During the past decades there has been a great deal of research to further understand the various nonlinear effects of these collective interactions. In laser-plasma interaction, the acceleration of fast electrons by Raman scattering, the nonlinear competition between Raman and Brillouin scattering, and the effects of filamentation on Raman scattering are but a few examples of these novel nonlinear effects. The reader is referred to Refs. 1 and 2 for a brief summary of these interesting advances.

Another area of the active research is the interaction of ultra high-power lasers with plasmas in which the oscillating velocity of the laser is approaching the speed of light. In this case, many relativistic nonlinear effects, such as the relativistic mass reduction and the corresponding change in the plasma frequency, become important, and fascinating physics has been revealed. Related to this is the super-short-pulse laser with sub-pico-second pulse durations, which also opens up a whole new area of investigations. The reader is referred to Refs. 3–5 for some examples of these studies. There has also been significant progress in our understanding of nonlinear Langmuir waves and the plasma turbulence associated with the Langmuir wave collapse and caviton formation, with application to ionospheric heating by powerful radar.[6] A number of interesting review articles can also be found in these proceedings on large-amplitude waves and fields in plasma.[6]

Intense electromagnetic wave interaction with plasma and charged particle beams will continue to be a most challenging and exciting field of study. We hope this book has, in some small way, aided the reader in acquiring some insight and understanding of this subject, as well as some simple mathematical tools for analyzing related problems of interest to him.[7]

References

1. H. A. Baldis, E. M. Campbell and W. L. Kruer, "Physics of laser-plasmas, in *Handbook of Plasma Physics*", Vol. 3, chap. 9 (North Holland, 1991).
2. T. W. Johnson, "Challenge to our understanding of scattering of laser plasmas", in *Inertial Confinement Fusion*, (American Institute of Physics, 1992), and references therein.
3. G. Mourou and D. Umstadter, *Phys. Fludis* **B4**, 2315 (1992).
4. P. Sprangle, E. Esarey and A. Ting, *Phys. Rev. Lett.* **64**, 2011 (1990).
5. S. C. Wilks, *Phys. Fluids* **B5**, 2603 (1993).
6. D. DuBois, H. A. Rose and D. Rusell, *Physica Scripta*, Vol. **T30**, 137–158 (1990).
7. R. Bingham *et al.* eds., "Large amplitude waves and fields in plasma", *Physica Scripta*, Vol. **T30** (1990).

www.ingramcontent.com/pod-product-compliance
Lightning Source LLC
Chambersburg PA
CBHW050642190326
41458CB00008B/2377
9789810215774